Microarray
Quality Control

Microarray Quality Control

Wei Zhang
Ilya Shmulevich
Jaakko Astola

A JOHN WILEY & SONS, INC., PUBLICATION

Library of Congress Cataloging-in-Publication Data:

Zhang, Wei, 1963 Oct. 9–
    Microarray quality control / Wei Zhang, Ilya Shmulevich, Jaakko
Astola.
        p. ; cm.
Includes bibliographical references and index.
    ISBN 0-471-45344-7 (cloth : alk. paper)
    1. DNA microarrays.
    [DNLM: 1. Oligonucleotide Array Sequence Analysis—standards. 2.
Quality Control. QZ 52 Z63m 2004]   I. Shmulevich, Ilya, 1969–   II.
Astola, Jaakko.   III. Title.
    QP624.5.D726 Z438 2004
    572.8'636—dc22   2003026922

Printed in the United States of America.

10 9 8 7 6 5 4 3 2 1

Contents

Foreword

Medical researchers and biologists have long recognized the tremendous genotypic and phenotypic heterogeneities in biological systems. Genotype–phenotype correlation studies have attempted to link characteristics as a basis for translating findings into practical applications, including individualized diagnosis and therapy. In the past, technology and methodology addressed only individual candidate genes or small groups of genes; global genotype–phenotype correlation was not possible. A major opportunity to achieve the goal of individualization appeared with the development of high-throughput genomic technologies, especially microarrays, with which thousands of genes can be evaluated simultaneously to generate a transcriptome profile or portrait for each individual. Now, the aim is to establish a correlation between these genetic portraits and specific phenotypes, such as propensity for certain diseases or the response of a disease to treatment. In recognition of the opportunity to fulfill this goal, many institutions and commercial entities have established genomics core facilities to produce microarrays and generate microarray data from biological materials. From 1998 to 2001, publications using microarray technologies have quadrupled each year, and microarray analysis has become a standard and nearly routine research method.

Interest in the field of genomics in the last few years has increasingly demonstrated that data generation and analysis are not trivial tasks. There are many steps where problems can occur, resulting in wasted efforts and squandered opportunities. The pitfalls begin with the choice of suitable biological materials and continue in microarray production, hybridization, imaging, and data analysis. The complexity of microarray technology poses a challenge. Most microarray users are biologists who are not trained in engineering, imaging, and data analysis. Many are not familiar with quality control/quality assurance in laboratory technology. On the other side of the lab bench, the personnel who deal with the nuts and bolts of array generation often lack background in biological studies and experimental design. As a consequence, there is a need for a quality control guide for researchers in the microarray field. This book specifically addresses quality control issues in the major steps of microarray applications.

The authors offer many useful suggestions based on their experiences and literature reports. They provide specific examples of problems that can occur when quality control measures are not taken. Thus, at this time, when the initial "hype" about microarrays has begun to settle into reality, attention to quality control continues to increase in importance. Microarray technology has developed into a necessary component in a biologist's toolbox. Therefore, the publication of this book is extremely timely for all researchers who produce, use, and analyze microarrays. As the methodology moves toward application to testing of individuals for decision making in clinical medicine, the stakes riding on quality control are even higher.

Stanley R. Hamilton, M.D.

Preface

From Nature's chain whatever link you strike,
Tenth or ten thousandth, breaks the chain alike.

—Alexander Pope (1688–1744)

Since their advent in 1995, microarrays have gained popularity at a tremendous rate and microarray core facilities have become commonplace in academic and industrial settings. The cost of setting up microarray facilities and performing experiments has been dropping steadily, making them more accessible to researchers. With this increased growth come new challenges, in particular those related to maintaining good quality control over the entire microarray experiment. Each stage of a microarray experiment presents its own pitfalls, and the aim of this book is to provide recommendations for identifying and preventing them.

This book focuses mainly on in-house microarray facilities and does not attempt to cover all aspects of quality control for commercially available arrays, such as Affymetrix chips. Nonetheless, many of the quality control issues, such as those dealing with biological sample handling and preparation as well as data analysis, will still be relevant and useful to those working with these technologies. At the end of each chapter we have provided a number of references relevant to the topics discussed in that chapter, in particular, to quality control issues. Not all of these references are cited in their respective chapters, but were selected by us, as they represent useful further reading material and provide many additional details.

As the title suggests, this book is about quality control and is not intended to be a "how-to" or tutorial book on microarrays. As such, it does not attempt to go into great depth in describing all phases of microarray production and analysis. For example, algorithms, statistical models, and so on, are developed only to the extent that they are necessary for the discussion, so as to make the book as self-contained as possible. In each chapter, some level of familiarity with the topics discussed therein is assumed. The book also relies on many examples for illustrating the concepts and issues. A number of other books on microarray production, experimentation, and data and image analysis, in addition to the journal articles listed in chapter references, may be of use to those wishing to study the relevant issues beyond the scope of this book. We feel that this book will be useful to both experienced microarray researchers as well as to those in the process of setting up microarray facilities. Our ultimate goal is to provide guidance on quality control so that each link in the microarray chain can be as strong as possible and the benefits of this exciting technology can be maximized.

We would like to express our gratitude to the following individuals for contributing to many of the quality control studies that formed the basis for some of the chapters: Latha Ramdas, David Cogdell, Limei Hu, Valerie Dunmire, Ellen Taylor, Jack Jia, Jing Wang, Ken Hess, and Kevin Coombes. We wish to thank Gregory Fuller, W. Fraser Symmans, and Lucian Chirieac for contributing the immunohistochemistry figures and related information as well as Sampsa Hautaniemi, Pekka Astola, Antti Lehmussola, Anna-Kaarina Järvinen, and colleagues in the Cancer Genomics Program at M. D. Anderson Cancer Center and Institute of Signal Processing at Tampere University of Technology for their kind help. We also would like to thank Ms. Beth Notzon and Ms. Marla Bordelon for editorial assistance. Finally, we would like to thank our editors, Luna Han and Angioline Loredo, for their hard work and encouragement, as well as the staff at Wiley for their outstanding editorial help.

1
Quality of Biological Samples

The increasing popularity of microarray analysis has been partially fueled by the frustration of biologists with limited technological tools they have had at their disposal for gaining a comprehensive understanding of complex biological problems. The completion of the Human Genome Project and the understanding it heralded threw into even greater relief the problems with the current technology. Although still not finally known, the number of genes in the human genome is estimated to exceed 40,000. These genes and their protein products determine biological phenotypes, both normal and abnormal. Traditional molecular biology methods have only allowed biologists to peek at the activities of a small number of genes that are associated with certain biological processes. In addition, although the reductionist method has enabled researchers to delineate a number of signal transduction pathways, it cannot yield a comprehensive picture of the systems under study. By allowing simultaneous measurement of the expression of thousands of genes, microarray technologies have marked a defining moment in biological research. Indeed, many publications have now shown the power of gene expression profiling in the diagnosis and prognosis of disease and the identification of gene targets. Microarray analysis is rapidly becoming a standard research tool.

It is therefore important for those outside the small pool of researchers who now do microarray analysis to know not only how it is done, but how to do it themselves. In general, a microarray experiment starts with the acquisition of biological materials from which RNA is isolated. For experiments involving established and relatively homogeneous cell lines, this is quite straightforward. However, for many experiments involving clinical tissues, the process is more complex and special attention must be paid to quality control. Because a central purpose of most microarray experiments is to map gene expression in biological samples, this chapter focuses on the quality control issues related to the subjects of experimentation—the tissues.

1.1   TISSUE ACQUISITION, HANDLING, AND STORAGE

In our expression microarray core facility, we have found that the best way to get good results from a microarray experiment is to start with RNA of exceptionally high quality. Isolating such RNA poses a significant problem, however, especially when the samples are taken directly from patients. However, this obstacle must be overcome because patient samples, as opposed to cell lines or animal models, tell the most physiologically relevant story.

Because RNA is not stable and degrades quickly if tissues are not handled and stored correctly, the method of acquiring, handling, and storing fresh tissues obtained during surgery for microarray analysis is critical to the success of the experiments. Some frequently encountered problems and possible consequences include:

Problem 1. Tissues are removed from patients by surgeons and placed on a counter, where they remain for hours before they are processed. Because the blood flow to the tissue is stopped, the cells start to undergo apoptosis, and RNase is released, which chews away at the RNA in the decaying cells.

Problem 2. Blood is drawn from patients and put into collection tubes, which are kept at room temperature for hours or overnight before they are transferred to the research laboratories. During this time, because the cells are under a different physiological condition and temperature, some cells begin to die, RNA begins to degrade, and gene expression begins to alter.

Problem 3. Tissues are cut into pieces and fixed in formalin. However, RNA is not stable in formalin and becomes degraded.

These problems highlight the importance of the active participation of clinical personnel in clinical research. To have the best tissues with good-quality RNA that accurately reflects the transcriptome in its physiological state, surgeons, pathologists, research nurses, and tissue bank personnel should know and strictly follow the correct tissue acquisition, handling, and storage procedures. First, tissues and blood samples should be kept at room temperature only for a brief time, preferably less than 30 minutes. After this, the tissues or blood cells should either be snap-frozen in liquid nitrogen or be put into solutions that inhibit RNase, with RNAlater—the most common RNase-inhibiting reagent. In the latter case, tissues should be cut into small pieces and immersed in the RNAlater, which can then penetrate the tissues and inhibit the RNase. Tissues can be maintained in this solution for a few weeks at 4°C. For long-term storage, however, tissues should be transferred to another tube and stored in a −80°C freezer. Alternatively, the tissues can be cut into small pieces, transferred to tubes that can be submerged in liquid nitrogen for snap-freezing, and stored in a −80°C freezer. Generally, tissues that are handled and stored in this way produce good-quality RNA. We prefer RNAlater to liquid nitrogen, however, because it is much more practical for research nurses in the operating room to put tissues into a tube prefilled with RNAlater than to prepare them in liquid nitrogen. It is also safer, in that liquid nitrogen poses a hazard in a patient-care setting.

The freezing solution is also important. For example, one that is used in the neuropathology laboratory at the University of Texas M. D. Anderson Cancer Center is the Tissue-Tek OCT Compound (Sakura/Tissue-Tek Company, Torrance, California), which is an inert whitish solution that helps maintain the morphology of the tissues during the cutting of tissues for slides. It consists of 10.24% polyvinyl alcohol, 4.26% polyethylene glycol, and 85.50% inert ingredients. Tissues stored in this solution should also be frozen as quickly as possible. OCT stands for optimal cutting temperature, and as the name implies, the solution was originally formulated to facilitate the cryostat sectioning of fresh frozen tissue (i.e., frozen section preparation). OCT has, however, also been proven to work well for tissue storage. It came to be used because, to make an intraoperative diagnosis, tissue can be embedded in OCT rather than paraffin (as is done in most hospitals), and then these OCT-embedded tissues can be frozen and stored. A second reason is that pathologists typically embed all of the extra tissue destined for the tumor bank in OCT solution because they want to be able to screen every block to determine, for example, whether the content is necrotic or pure, and the only way to cut sections for this is to embed the tissue in OCT. Such screening is not necessary with some tumors (e.g., meningiomas) that are clearly pure, even grossly. However, this would not be the case for gliomas because of the diffuse infiltration that may be present. Of course, one can flash-freeze tissue directly in liquid nitrogen, but then one cannot use such material to cut tissue sections. Some laboratories do freeze tissue, even glioma tissue, directly in liquid nitrogen and perform experiments with it, simply assuming that the tumors are pure, but assumption is a dangerous business in science because it can lead to erroneous conclusions, and therefore this practice is not recommended.

In the next section we discuss the importance of pathological evaluation. Before that, however, we describe those tissue collection and storage methods that yield the most optimal RNA quality. This is especially important for biopsy tissue and fine-needle aspirates, which are very limited in quantity.

In a study carried out in our laboratory, we compared the RNA extracted from tissue biopsy specimens that were prepared in four different ways: (1) snap-frozen in liquid nitrogen; (2) preserved in RNAlater; (3) preserved in RNAlater, fixed in ethanol, and embedded in OCT; and (4) preserved in RNAlater, fixed in ethanol, and embedded in paraffin. In this experiment, normal and tumor tissue samples were obtained in a standard clinical manner by surgical resection from seven patients with colon cancer. Each tissue sample was divided into four portions and subjected to one of the four study conditions, all according to standard hospital procedure: The paraffin-embedded samples were stored at room temperature until the RNA was extracted, whereas the other three were stored at $-80°C$ until the RNA was extracted.

Each RNA sample was analyzed with a spectrophotometer at 260, 280, and 230 nm to detect protein and polysaccharide contamination and determine the RNA concentration. RNA degradation and genomic DNA contamination were assessed in aliquots of the preparations by electrophoresis on a 1% denaturing agarose gel containing 1X MOPS (morpholinopropanesulfonic acid) buffer and 2% formaldehyde. Each sample was also loaded into an RNA LabChip (Caliper Technologies Corp., Mountain View, California) and analyzed on the Agilent 2100 bioanalyzer (Agilent Technologies, Palo Alto, California) using the RNA 6000 Nano Assay.

Two representative RNA samples were also used to hybridize to a cDNA microarray containing 4800 features produced in-house. Good microarray results were obtained for all but the samples embedded in paraffin, indicating that the RNA isolated from the biopsy samples prepared under the other three conditions was of good quality for microarray analysis. Nonetheless, for the reasons given earlier, we prefer RNAlater to liquid nitrogen as a processing agent.

Clinical samples obtained by fine-needle aspiration (FNA) pose a unique challenge, however. Unlike biopsy specimens, in which cells are in a lump of tissues, tissues obtained by FNA are often broken up, and when they are immersed in RNAlater, the cells are often dispersed. FNA samples from tumors are particularly useful, however, because they represent a relatively pure population of tumor cells that is relatively free of the contaminating normal cells found in a surgical resection specimen. This purity is particularly important when one considers the subtle changes that may occur at the beginning of tumorigenesis. For example, contaminating normal cells could obscure a small but important change in a tumor's gene expression pattern. FNA specimens are also preferable because FNA is much safer and less invasive than biopsy or surgery. FNA could also enable repeated aspirates to be collected from the same tumor so that the progression of the disease and the tumor's response to therapy could be monitored. It is therefore important to be able to use FNA samples for microarray analyses because of the valuable potential they hold for gaining insight into the molecular details and prognostic indicators for certain types of cancer for which there is a limited volume of relatively pure tumor cells. Indeed, such clinical samples can be used with high efficiency in high-density expression microarray analyses. For this reason, it is essential to retrieve as much high-quality genetic material as possible from each FNA sample.

We thus also evaluated RNA quality and quantity in FNA specimens and found that because the cells are partially lysed, up to 50% of the RNA in some specimens is released from the cells into the RNAlater solution and thus lost, which is quite significant for FNA samples. In most protocols involving the use of RNAlater, tissues are separated from the RNAlater solution by centrifugation and then the pellet is exposed to an RNA isolation solution. To minimize the loss of RNA, we suggest keeping the RNAlater solution portion and also recovering the RNA from the RNAlater solution. To determine the feasibility of doing this, we used RNA isolated from the RNAlater supernatant and RNA isolated from the FNA pellet for microarrays and found that the results from both were very concordant. Thus, excessive RNAlater should not be added to the FNA tissue at the acquisition step. Two hundred microliters should be sufficient. FNA samples, therefore, yield sufficient RNA of adequate quality for microarray analysis, qualities that together allow for valuable high-density analysis.

1.2 PATHOLOGICAL EVALUATION

A common problem with microarray experiments involving the use of clinical samples is that a pathological evaluation of the tissues is not done before the experiments. This can lead to flawed results. We will illustrate this problem using two experiences we have had in our laboratory. The first example comes from a genomic study of sarcomas. Sarcoma is a cancer of soft tissues. Leiomyosarcoma is a type of sarcoma of muscle origin that often grows in the gastrointestinal tract for a long time before being discovered. Some of these weigh 20 pounds (9 kg) or more when they are finally removed during surgery. To remove most of the tumor that may have infiltrated into neighboring normal tissues, surrounding normal tissues, including the stomach, are also routinely removed. The tumors are then cut into small pieces for tissue banking, some of which are used for research purposes. The normal tissues removed during surgery are used as controls. It is, therefore, not difficult to envision that sometimes the pieces labeled "tumor" may actually contain a large fraction of normal tissue. In the absence of a pathological evaluation of the tissues, the true nature of the tissue would not be known. Similarly, normal stomach tissues can consist of either muscle cells or epithelial cells, depending on where the cut is made. Since muscle tissues are the corresponding normal tissues for leiomyosarcomas, a tissue sample composed mainly of epithelial cells would not be the correct control material for a leiomyosarcoma. Pathological evaluation would ensure that the correct normal tissues are used as the control material.

The second example comes from a study in which we were using specimens from breast cancer lymph node metastases. Out of the 18 samples removed at surgery, pathological examination revealed that eight had infiltrating lymphocytes that constituted more than 50% of the cell population. Thus, these samples were not used for the subsequent microarray study. If we had not performed this evaluation and used those tissues in the study, although the number of specimens in the study would have been higher, the results would have been highly confounded because of the presence of a large percentage of nontumor cells.

A third problem that can arise is errors in pathological classification and database tracking. It is commonly known by pathologists that many tumors have atypical morphologies and categorizing these cases sometimes involves guesswork. Different pathologists may even disagree on a diagnosis. Honest mistakes can also occur and sometimes records get mixed up. Thus, it is always prudent to reevaluate the tissues to be used in a molecular study to confirm that they are what they are supposed to be. There are also stories of highly regarded molecular classification studies using microarray analysis that turn out to be flawed due to problems of misclassification resulting from misdiagnosis during the pathological reevaluation of the material.

When molecular and genomic biologists first enter the world of pathology, they typically are amazed at the complexity of tissues—a world far different from that of pure tumor cells in a culture dish, which laboratory researchers typically enjoy. What pathologists see is quite a frustratingly heterogeneous picture of various cell populations. We describe some of these real-world experiences here.

1.3   TISSUE HETEROGENEITY AND LASER CAPTURE MICRODISSECTION

The view of cancer researchers used to working with cancer cell lines can be very simplistic. They often view cancers as being a large group of cancer cells growing together. However, this is far from reality. *In vivo* tissues, whether normal or diseased, are very heterogeneous. For example, in blood, there are cells of different lineages: the highly specialized red blood cells, which transport oxygen to the body; the lymphoid cells (B and T cells), which are important to the immune response; and the myeloid cells (monocytes, granulocytes, and macrophages), which are important to the inflammatory response, among other functions. In breast tissues, there are ductal epithelial cells, stromal cells such as fibroblasts, endothelial cells that make up the blood vessels, fat cells, and immune cells such as lymphocytes. Even in the *immune-privileged brain*, there are neurons as well as several different types of glial cells, such as astrocytes, oligodendrocytes, and microglial cells. Because of the heterogeneous nature of normal cells, therefore, it is often difficult to use normal tissues as "normal" control for comparison studies. For example, unless the normal astrocytes are microdissected, normal brain tissues, which contain many neurons, are not a suitable normal control material for astrocytomas. In leukemia studies, to serve as a normal control for B-cell malignancies, normal cell populations such as B cells can be sorted out by flow cytometry [or fluorescent-activated cell sorting (FACS)] using B-cell-specific surface markers. Cell surface markers in combination with FACS sorting can also be used to purify many other cell types, such as endothelial cells.

In cancer, a major area of research in which microarray studies are used, the tissues obtained at surgery can also be highly heterogeneous, although the core of the tumors may be rich in cancer cells. These heterogeneities are manifested at several levels. The first level of heterogeneity may cause the most artifacts if quality control is not observed. Because most tissues are obtained during surgical removal and to ensure a near-complete resection of the tumor and invasive tumor cells, surrounding normal tissues are also routinely removed, most tumor tissue specimens have contaminating normal tissues. A particular tissue block used for research may also have no tumor tissue, or different blocks may have different percentages of tumor cells. From this, one can see that obtaining an accurate tumor-specific gene expression signature strongly depends on the quality of the specimens. It is therefore extremely important to have the piece of tissue evaluated by a pathologist before processing it for microarray experiments. In such an evaluation, pathologists normally cut a section from the top and the bottom of the tissue specimen and evaluate the tumor makeup after routine H&E staining. Some trimming may then be done to remove contaminating normal tissues.

The second level of heterogeneity is biological. For example, in the case of a human glioma, pathological evaluation of the tissues often reveals mixed morphological features. As shown in Figure 1.1, panels A to D, four microscopic fields from the same glioblastoma show small cell (A), spindle cell (B), giant cell (C), and epithelioid (D) differentiation. In other tissues, all these features may be intermingled. Different tissues can also have different amounts of vascularization. In particular, low-grade tumors typically have low levels of angiogenesis, and high-grade tumors have high levels of angiogenesis. Therefore, a difference in the gene expression signature between different tumor tissues may imply different things, one of which could be a difference in the amount of endothelial cells in the tumor.

Even when tissues look homogeneous in their morphology, different cells in the tissues may not behave the same at the molecular level. This is best illustrated by immunohistochemistry studies using different molecular markers. To use astrocytomas again as an example, two different areas of the same oligodendroglioma show strikingly different proliferation activities, indicated by staining of MIB-1 marker (Figure 1.1, panels E and F). Another example is that the glial cell marker glial fibrillary acidic protein (GFAP) can have variable levels of expression in different areas of one glioblastoma (Figure 1.1, panels G and H). This is also the case for many other markers in different subsets of tumor cells in some localized region.

A further consideration is that tumor tissues such as colon and breast cancers are often significantly intermingled with stromal cells and infiltrated with inflammatory cells and lymphocytes. As shown in panels A to H in Figure

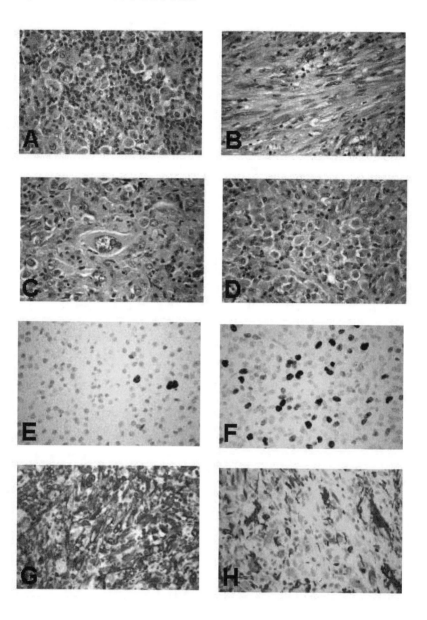

Fig. 1.1 Phenotypic, proliferative, and molecular heterogeneity in brain tumors. Phenotypic heterogeneity (A–D): Four microscopic fields from the same glioblastoma show small cell (A), spindle cell (B), giant cell (C), and epithelioid (D) differentiation. (H&E, 200×.) Proliferative heterogeneity (E–F): Two different areas of the same oligodendroglioma show strikingly different proliferation activities. (MIB-1 immunocytochemistry; 200×.) Molecular heterogeneity (G–H): Two areas of one glioblastoma illustrate variable GFAP gene expression. (GFAP immunocytochemistry; 200×). (This figure was generously provided by Dr. Gregory N. Fuller, The University of Texas M. D. Anderson Cancer Center.) See insert for a color representation of this figure.

1.2, highly morphologically different colon cancer cells display invasion into stromal cells and adipose tissues and are infiltrated with lymphocytes. Similarly, in breast cancer tissues, different parts of the same tumor may have very different cell populations. Some tumor specimens are composed mostly of cancer cells (Figure 1.3, panel A), whereas other tumor specimens have less than 50% of tumor cells (Figure 1.3, panel B). Different breast metastatic tumors in the lymph nodes can have drastically different amounts of infiltrating lymphocytes. Therefore, a comparison of the gene expression of different breast cancer tissues could ultimately show only the difference between breast cancer cells and lymphocytes should tumor-rich and lymphocyte-rich tissues be compared.

These examples point out the heterogeneous nature of the tissues we study and underscore the impact that different populations of cells in a tissue can have on the final microarray results, which only reflect the population average. Obviously, pathological evaluation is an important quality control step when selecting suitable tissues for a particular study. Laser capture microdissection technology was developed as a means of further overcoming this problem. In this technology, cells of interest are dissected from the tissues for study. However, although this technology has been extremely successful in isolating cells from paraffin-embedded tissues for DNA-based assays, frozen tissues have to be used for RNA-based assays such as microarrays, and a key limiting factor is isolating a sufficient amount of cells so that there is enough RNA for the microarray study. Nonetheless, with the development of amplification protocols for microarray analyses (discussed in Chapter 3), this technology is now more feasible and its popularity will probably surge in the coming years.

A drawback of the selection of cells by microdissection, however, is that it is based on subjective standards such as cell morphology, but morphologically similar cells may not be genotypically similar. To obtain an accurate picture, therefore, one may need to perform single cell-based assays. However, this then takes us away from the systems point of view of biology. In addition, this may not be practical because we do not know how many single cells we need to examine before we can draw a conclusion about the tissues. Therefore, there needs to be a balance between the reductionist and systems approaches. In other words, we need to remove those things that will interfere with the systems under study or the goals of the study, but not those things that are part of the systems. For example, angiogenesis and inflammation are part of cancer systems. Indeed, recent studies have shown that tumor cells often recruit inflammatory cells and make them coconspirators in tumorigenesis by inducing them to secrete proteins such as metalloproteinase, which facilitate tumor cell invasion and metastasis. Obviously, microdissection would remove these important cancer factors from the systems.

The intimate cross-talk between cancer cells and noncancer cells in cancer tissues is also illustrated by one of our recent studies (Kobayashi *et al.*, 2003). Diffuse large B-cell lymphoma (DLBCL) is a very heterogeneous disease with different surface markers expressed in the cells of different patients. One

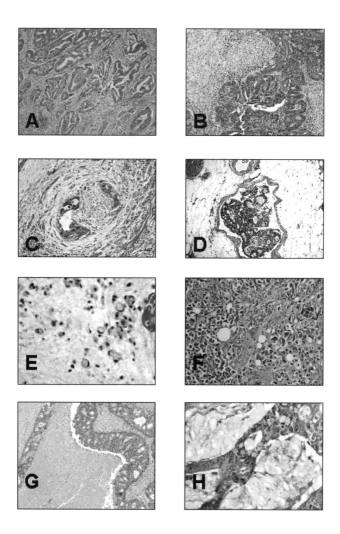

Fig. 1.2 Phenotypic heterogeneity in colon tumors. Moderately differentiated adenocarcinoma of colon (A–D): Elongated malignant glands arranged in a corkscrew pattern invade into the stroma (A). A different tumor displays a cribriform pattern and a prominent periglandular lymphocytic infiltrate (B). (H&E, 200×.) Same tumor focally invades the perineurium (C) and is present in the vascular space (D). (H&E, 400×.) Poorly differentiated adenocarcinoma of colon (E–F): Discohesive signet-ring cells floating in pools of mucin (E). (H&E, 400×.) Highly anaplastic tumor with pleomorphic cells, invading into the pericolonic adipose tissue (F). (H&E, 200×.) Adenocarcinoma of colon with ribbonlike pattern (G–H): Enlarged, dilated malignant glands with central necrosis (G). (H&E, 200×.) Mucinous tumor with pleomorphic cells and abundant extracellular mucin (H). (H&E, 400×.) (This figure was generously provided by Dr. Lucian Chirieac, The University of Texas M. D. Anderson Cancer Center.) See insert for a color representation of this figure.

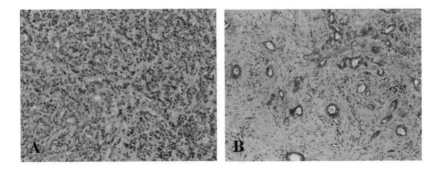

Fig. 1.3 The histopathological appearance is variable among different invasive breast cancers. Some cancers consist of an almost pure population of breast cancer cells (A). Most cancers contain a significant component of desmoplastic stroma (B) that is active in angiogenesis, fibrosis, myocontractility, and immunity. (This figure was generously provided by Dr. Fraser Symmans, The University of Texas M. D. Anderson Cancer Center.) See insert for a color representation of this figure.

group (10% of patients with DLBCL) is positive for the CD5 marker and has a poorer overall prognosis than patients with CD5-negative DLBCL. We carried out a microarray study to characterize the gene expression activities of CD5-positive and CD5-negative DLBCL. Among the differentially expressed genes were integrin beta 1 and CD36. Immunohistochemical staining then showed that integrin beta 1 indeed is expressed in most CD5-positive DLBCLs, which is consistent with the greater invasiveness of this subtype of DLBCL. The staining for CD36 also showed some striking patterns. CD36 was expressed in CD5-positive DLBCLs, but only in the endothelial cells of large vessels. In contrast, CD36 was not expressed in CD5-negative DLBCLs, although the large vessels were present (Figure 1.4). This is a good example of normal cells in different cancer tissues expressing different genes, which very likely contributes to the different biological behavior of the tumors. Microdissection of only the tumor cells for study would have missed this potentially important information. One solution, of course, would be to microdissect the vessel cells to obtain vessel-specific biological information as well.

1.4 SUMMARY

It is important to appreciate tissue heterogeneity and its effect on the results of microarray analyses. As a quality control step, it is worthwhile to have the active participation of pathologists to evaluate the cellular composition of the tissues under study. Depending on the goals of a study, one then has to decide how much tissue and cell purification to do, either by coarse-

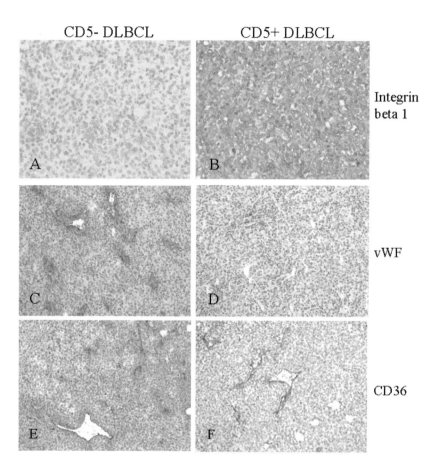

Fig. 1.4 Immunohistochemical staining of lymphoma tissues for integrin beta 1, vWF, and CD36. Integrin beta 1 was weakly stained for CD5-negative DLBCL (A) and strongly stained for CD5-positive DLBCL (B). vWF was stained strongly in the vascular cells of both CD5-negative and CD5-positive DLBCL (C and D). CD36 was not stained in CD5-negative DLBCL (E) and stained highly in the vascular cells of CD5-positive DLBCL (F). See insert for a color representation of this figure.

scale trimming or laser capture microdissection. The results of population-average studies eventually need to be validated by cell-based assays, such as immunohistochemistry analyses to gain spatial information on gene activities.

REFERENCES

1. Assersohn L, Gangi L, Zhao Y, Dowsett M, Simon R, Powles TJ, Liu E. (2002) The feasibility of using fine needle aspiration from primary breast cancers for cDNA microarray analyses. *Clin Cancer Res* 8:794–801.

2. Ball HM, Hupp TR, Ziyaie D, Purdie CA, Kernohan NM, Thompson AM. (2001) Differential p53 protein expression in breast cancer fine needle aspirates: the potential for in vivo monitoring. *Br J Cancer* 85:1102–1105.

3. Barroca H, Carvalho JL, da Costa MJ, Cirnes L, Seruca R, Schmitt FC. (2001) Detection of N-myc amplification in neuroblastomas using Southern blotting on fine needle aspirates. *Acta Cytol* 45:169–172.

4. Chang J, Ormerod M, Powles TJ, Allred DC, Ashley SE, Dowsett M. (2000) Apoptosis and proliferation as predictors of chemotherapy response in patients with breast carcinoma. *Cancer* 89:2145–2152.

5. Chehade JM, Silverberg AB, Kim J, Case C, Mooradian AD. (2001) Role of repeated fine-needle aspiration of thyroid nodules with benign cytologic features. *Endocr Pract* 7:237–243.

6. Chiang MK, Melton DA. (2003) Single-cell transcript analysis of pancreas development. *Dev Cell* 3:383–93.

7. DeRisi J, Penland L, Brown PO, Bittner ML, Meltzer PS, Ray M, Chen Y, Su YA, Trent JM. (1996) Use of a cDNA microarray to analyse gene expression patterns in human cancer. *Nat Genet* 4:457–60.

8. Dubois NA, Kolpack LC, Wang R, Azizkhan RG, Bautch VL. (1991) Isolation and characterization of an established endothelial cell line from transgenic mouse hemangiomas. *Exp Cell Res* 2:302–13.

9. Dunmire V, Wu C, Symmans WF, Zhang W. (2002) Increased yield of total RNA from fine-needle aspirates for use in expression microarray analysis. *Biotechniques* 4:890–2, 894, 896.

10. Emmert-Buck MR, Bonner RF, Smith PD, Chuaqui RF, Zhuang Z, Goldstein SR, Weiss RA, Liotta LA. (1996) Laser capture microdissection. *Science* 274:998–1001.

11. Fearon ER, Vogelstein B. (1990) A genetic model for colorectal tumorigenesis. *Cell* 61:759–767.

12. Fink L, Seeger W, Ermert L, Hanze J, Stahl U, Grimminger F, Kummer W, Bohle RM. (1998) Real-time quantitative RT-PCR after laser-assisted cell picking. *Nat Med* 4:1329–33.

13. Florell SR, Coffin CM, Holden JA, Zimmermann JW, Gerwels JW, Summers BK, Jones DA, Leachman SA. (2001) Preservation of RNA for functional genomic studies: a multidisciplinary tumor bank protocol. *Mod Pathol* 14:116–128.

14. Galindo LM, Garcia FU, Hanau CA, Lessin SR, Jhala N, Bigler RD, Vonderheid EC. (2000) Fine-needle aspiration biopsy in the evaluation of lymphadenopathy associated with cutaneous T-cell lymphoma (mycosis fungoides/Sezary syndrome). *Am J Clin Pathol* 113:865–871.

15. Grotzer MA, Patti R, Geoerger B, Eggert A, Chou TT, Phillips PC. (2000) Biological stability of RNA isolated from RNAlater-treated brain tumor and neuroblastoma xenografts. *Med Pediatr Oncol* 4:438–42.

16. Gudlaugsdottir S, Sigurdardottir V, Snorradottir M, Jonasson JG, Ogmundsdottir H, Eyfjord JE. (2000) P53 mutations analysis in benign and malignant breast lesions: using needle rinses from fine-needle aspirations. *Diagn Cytopathol* 22:268–274.

17. Hanahan D, Weinberg RA. (2000) The hallmarks of cancer. *Cell* 100:57–70.

18. Kimler BF, Fabian CJ, Wallace DD. (2000) Breast cancer chemoprevention trials using the fine-needle aspiration model. *J Cell Biochem* 77:7–12.

19. Kobayashi T, Yamaguchi M, Kim S, Morikawa J, Ogawa S, Ueno S, Suh E, Dougherty E, Shmulevich I, Shiku H, Zhang W. (2003) Microarray reveals differences in both tumors and vascular specific gene expression in de novo CD5+ and CD5– diffuse large B-cell lymphomas. *Cancer Res* 63:60–6.

20. Maitra A, Wistuba II, Virmani AK, Sakaguchi M, Park I, Stucky A, Milchgrub S, Gibbons D, Minna JD, Gazdar AF. (1999) Enrichment of epithelial cells for molecular studies. *Nat Med* 5:459–63.

21. Makris A, Powles TJ, Allred DC, Ashley SE, Trott PA, Ormerod MG, Titley JC, Dowsett M. (1999) Quantitative changes in cytological molecular markers during primary medical treatment of breast cancer: a pilot study. *Breast Cancer Res Treat* 53:51–59.

22. *Nature,* April 24, 2003.

23. Pabon C, Modrusan Z, Ruvolo MV, Coleman IM, Daniel S, Yue H, Arnold, LJ, Jr. (2001) Optimized T7 amplification system for microarray analysis. *Biotechniques* 31:874–879.

24. Panaro NJ, Yuen PK, Sakazume T, Fortina P, Kricka LJ, Wilding P. (2000) Evaluation of DNA fragment sizing and quantification by the agilent 2100 bioanalyzer. *Clin Chem* 46:1851–1853.

25. Pusztai L, Ayers M, Stec J, Clark E, Hess K, Stivers D, Damokosh A, Sneige N, Buchholz TA, Esteva FJ, Arun B, Cristofanilli M, Booser D, Rosales M, Valero V, Adams C, Hortobagyi GN, Symmans WF. (2003) Gene expression profiles obtained from fine-needle aspirations of breast cancer reliably identify routine prognostic markers and reveal large-scale molecular differences between estrogen-negative and estrogen-positive tumors. *Clin Cancer Res* 9:2406–2415.

26. Sato N, Yamashita H, Kozaki N, Watanabe Y, Ohtsuka T, Kuroki S, Nakafusa Y, Ota M, Chijiiwa K, Tanaka M. (1996) Granulomatous mastitis diagnosed and followed up by fine-needle aspiration cytology, and successfully treated by corticosteroid therapy: report of a case. *Surg Today* 26:730–733.

27. *Science,* April 11, 2003.

28. Shmulevich I, Hunt K, El-Naggar A, Taylor E, Ramdas L, Laborde P, Hess KR, Pollock R, Zhang W. (2002) Tumor specific gene expression profiles in human leiomyosarcoma: an evaluation of intratumor heterogeneity. *Cancer* 94:2069–2075.

29. Sotiriou C, Powles TJ, Dowsett M, Jazaeri AA, Feldman AL, Assersohn L, Gadisetti C, Libutti SK, Liu ET. (2002) Gene expression profiles derived from fine needle aspiration correlate with response to systemic chemotherapy in breast cancer. *Breast Cancer Res* 4(3):R3.

30. Symmans WF, Volm MD, Shapiro RL, Perkins AB, Kim AY, Demaria S, Yee HT, McMullen H, Oratz R, Klein P, Formenti SC, Muggia F. (2000) Paclitaxel-induced apoptosis and mitotic arrest assessed by serial fine-needle aspiration: implications for early prediction of breast cancer response to neoadjuvant treatment. *Clin Cancer Res* 6:4610–4617.

31. Symmans WF, Ayers M, Clark EA, Stec J, Hess KR, Sneige N, Buchholz TA, Krishnamurthy S, Ibrahim NK, Buzdar AU, Theriault RL, Rosales MF, Thomas ES, Gwyn KM, Green MC, Syed AR, Hortobagyi GN, Pusztai L. (2003) Total RNA yield and microarray gene expression profiles from fine-needle aspiration biopsy and core-needle biopsy samples of breast carcinoma. *Cancer* 97:2960–71.

32. van de Vijver MJ, He YD, van't Veer LJ, Dai H, Hart AA, Voskuil DW, Schreiber GJ, Peterse JL, Roberts C, Marton MJ, Parrish M, Atsma D, Witteveen A, Glas A, Delahaye L, van der Velde T, Bartelink H, Rodenhuis S, Rutgers ET, Friend SH, Bernards R. (2002) A gene-expression signature as a predictor of survival in breast cancer. *N Engl J Med* 347:1999–2009.

33. Van Gelder RN, von Zastrow ME, Yool A, Dement WC, Barchas JD, Eberwine JH. (1990) Amplified RNA synthesized from limited quantities of heterogeneous cDNA. *Proc Natl Acad Sci USA* 87:1663–1667.

34. van 't Veer LJ, Dai H, van de Vijver MJ, He YD, Hart AA, Mao M, Peterse HL, van der Kooy K, Marton MJ, Witteveen AT, Schreiber GJ, Kerkhoven RM, Roberts C, Linsley PS, Bernards R, Friend SH. (2002) Gene expression profiling predicts clinical outcome of breast cancer. *Nature* 415:530–6.

35. Walch A, Specht K, Smida J, Aubele M, Zitzelsberger H, Hofler H, Werner M. (2001) Tissue microdissection techniques in quantitative genome and gene expression analyses. *Histochem Cell Biol* 115:269–76.

36. Zhuang Z, Bertheau P, Emmert-Buck MR, Liotta LA, Gnarra J, Linehan WM, Lubensky IA. (1995) A microdissection technique for archival DNA analysis of specific cell populations in lesions < 1 mm in size. *Am J Pathol* 146:620–5.

2

Microarray Production: Quality of DNA and Printing

In most microarray facilities, there are two types of microarrays that are generally produced: cDNA microarrays, in which the PCR products of cDNA clones are printed, and long-oligonucleotide (oligo) arrays, in which oligos of a certain length are printed. Because all subsequent experiments and data generation rely on the quality of the microarray slides, their production is critically important and requires the maintenance of rigorous quality control. In this chapter we describe some of the common problems that microarray facilities can encounter in the production of microarrays and the quality control measures that can be taken to avoid these problems.

There are also several key choices to be made when producing microarrays. These include deciding whether to use cDNA or oligos, choosing the oligo length, and deciding what coating to put on slides. Other choices involve the source of the clone libraries, the types of the printer, and other hardware. Although it is not our intention to compare products in this book, we will discuss factors that researchers should consider when purchasing products.

2.1 QUALITY CONTROL FOR cDNA PROBES

Most microarray facilities purchase cDNA clone libraries from Research Genetics (Huntsville, Alabama) or directly from the Integrated Molecular Analysis of Genomes and their Expression (IMAGE) consortium. Some facilities use cDNA clone libraries produced in house, such as subtracted libraries, for specific projects. The flexibility afforded by the latter is one of the major benefits of in-house microarray facilities. Further, an obvious advantage of using a cDNA clone library is that the identity of the clones in the library does not need to be known before microarrays can be generated, as long as the clones are tracked. Therefore, microarray experiments can be performed before the sequencing process is completed for some species (such as cow and dog). Microarray experiments can also be performed to identify interesting genes first, followed later by sequencing. This is especially economical for screening purposes.

Most cDNA libraries come in a stack of 96-well plates in which bacteria harboring the cDNA in plasmid are stored. This format allows easy duplication and expansion of the libraries for distribution. To produce a microarray, thousands of cDNA probes are first obtained by polymerase chain reaction (PCR) amplification using a common pair of primers located on the vector used for construction of a bacterial clone library. Some researchers first prepare plasmid DNAs from the clone library and then use them as templates for PCR; however, most laboratories find this step unnecessary. Routinely, the library is first duplicated and an aliquot of the bacterial culture (5 μL) is used directly as the template in a PCR assay. High-throughput PCR in a 96-well format is normally set up either using a robotic liquid-handling system or manually using a multichannel pipettor. After PCR, the PCR products are purified and then electrophoresed on a gel to evaluate the PCR production and the size of the product. Those samples that produce sufficient amounts of the product of the correct size can then be cherry-picked by the liquid-handling system for final DNA source plates for sequence validation and microarray printing. Some facilities use PCR products directly for printing without purification (Diehl *et al.*, 2002). The advantage of this is that less DNA material is lost as the result of purification and thus more microarrays can be produced with the PCR products. A drawback to this is that any impurity in the reaction (primers and enzymes) may adversely affect hybridization and cause high background interference.

High-throughput operations that involve multiple steps in sample handling and processing, such as microarray production and library construction, can be subject to error. For example, the sequences of all cDNA clones are supposed to have been verified when the clones were added to the IMAGE consortium and should be correct. However, mistakes could occur and be propagated when the collections are repeatedly duplicated, amplified, and transferred. Indeed, the IMAGE consortium indicated that the library had a sequence error rate of 12% (http://image.llnl.gov/image/qc/bin/display_error_rates); in fact, an evaluation of 1189 IMAGE cDNA clones showed a sequencing error rate as high as 38% (Halgren *et al.*, 2001). Those IMAGE clones sent to Research Genetics were also resequenced with a database created describing the gene identities, but once again, the sequence-verified cDNA library was repeatedly amplified and distributed. Thus, errors could occur and be propagated at both the vendor's and user's end.

The frequencies of errors may vary depending on the method of liquid transfer. Early on in our microarray facility, we used a multichannel pipettor for this purpose. To avoid a transfer mistake, we used a two-person rotation approach in which one person performed the pipetting and one person watched over his or her shoulder. However, even with this safeguard, human concentration tended to lessen with time, leading to mistakes. Therefore, we quickly adopted the use of a liquid-handling robotic system to handle all liquid transfer. We believe that this system is less prone to errors and thus is a very important quality control investment for all microarray facilities. There are many different robotic systems on the market to choose from but we have been using a robotic system from Tecan (Raleigh, North Carolina) with satisfying results.

Once this quality control liquid-handling step was in place, we tested whether a sequence verification step late in the production process was a crucial quality control step for microarrays made of annotated genes, especially when the gene names are provided to the end users. A sequence-verified cDNA clone library was purchased from Research Genetics. Clones from 32 plates in a 96-well format were transferred from the frozen bacterial stock to culture plates. These clones were then grown at 37°C for 1–2 days. A liquid-handling robot was used to set up the PCRs with 5 μL of the bacterial culture. After thermocycling, the robot was used again for PCR purification with PCR purification plates (Millipore, Bedford, Massachusetts). The PCR products were run on 1% agarose gels to check for purity and to verify size.

The master plates of cDNA clones were then cycle-sequenced with fluorescent dye-labeled terminators using an ABI Prism BigDye Terminator Cycle Sequencing Kit with AmpliTaq DNA polymerase. The sequence was initially analyzed using the Sequence Analysis software v3.6 (PE Applied Biosystems). The nucleotide sequence data from each reaction were individually copied from the 3700 Sequence Analysis software into a text file. The files were then converted to a FASTA format in which each description line identified the clone location provided by Research Genetics.

The sequence data were then analyzed using an in-house sequence verification program, Rocket:

http://www3.mdanderson.org/depts/cancergenomics/software.html

The default assumption built into this program was that the information supplied by Research Genetics was correct. The algorithm was structured to verify correctness quickly, with more complicated searches done only if the initial tests failed.

The program first performed a BLAST2 search of the sequence from our text file against the GenBank accession number provided by Research Genetics, using the BLOSUM62 clustered scoring matrix and the default parameter settings at the National Center for Biotechnology Information Web site (http://www.ncbi.nlm.nih.gov/blast/bl2seq/bl2.html). Because these parameters did not require exact matches, the program overlooked minor dis-

Gel Analysis	Number of Clones	Percent
Correct size	1918	63
Incorrect size	211	7
Multiple bands	408	13
Low or no yield	534	17
Total clones amplified (32 plates)	3072	100

Table 2.1   Gel analysis of all clones before assembling into final plate format.

crepancies in sequencing. If the BLAST2 search showed that the sequences matched, the result was written to a log file and the program moved on to consider another clone.

If the BLAST2 search failed to show a match, then the program used our sequence to perform a full BLAST search of the nonredundant human database to find the ten best matches. The algorithm found the current UniGene cluster numbers for both the clone information provided by Research Genetics and for each of the top ten matches. If any of the UniGene cluster numbers agreed, then the result was logged as a match between the sequence and the accession number.

If the combined BLAST/UniGene procedure failed to match our sequence with the accession number, then an error was logged and the algorithm searched for contamination from other clones supplied by Research Genetics . This was done by searching with the top ten UniGene cluster matches for our sequence against other UniGene clusters in the information supplied by the vendor. If this identified a match, then a confirmation of the result was attempted by performing another BLAST2 search of our sequence against the matching accession number. If this identified a match, then it was logged as a match to a contaminating clone. Finally, if all attempts at identifying the sequence inside Research Genetics' data failed, then an error was logged and the accession number of the best match from the full BLAST search was used to annotate the clone.

The PCR products amplified from the cDNA clones were first analyzed on a 1% agarose gel, with each lane corresponding to one clone. Most lanes (70%) showed a single strong band, whereas some lanes showed multiple bands, a low yield, or no amplification. Each single-banded product was compared with the Research Genetics database to check clone insert size (Table 2.1).

Most showed the correct-size DNA (63%), but a small percentage (7%) showed the incorrect-size DNA. This helped detect potential gross errors, such as plate inversions. Many times, the lanes containing more than one band had a band of the correct size and a band corresponding to the size of a neighboring well, suggesting cross-contamination. Several original bacterial stocks that showed multiple bands after PCR were streaked onto new Luria Broth (LB) bacterial culture plates, and a number of individual colonies were then picked and amplified by PCR. Gel electrophoresis showed that two particular populations of cells often existed, suggesting that cross-contamination occurred during library duplication.

To determine whether the DNA size was a reliable indicator for the clones, we picked for sequencing a similar number of clones with both a correct-size DNA ($n = 107$) and an incorrect size ($n = 121$). After sequencing they were searched by BLAST. As shown in Table 2.2, even among the correct-size DNA group, 15% did not match the genes named in the database. Interestingly, among the incorrect-size DNA group, 22% actually matched the genes named in the vendor's database. This observation led us to two conclusions. First, there can be errors in the cDNA size shown in the vendor database, and second, to guarantee that the names match the cDNAs on our array, a final sequencing step is needed.

Correct Size (107)		
Correct sequence	91	85%
Incorrect sequence	16	15%
Incorrect Size (121)		
Correct sequence	24	22%
Incorrect sequence	97	78%

Table 2.2 Quality control sequence results.

After the final cDNA for microarray printing contained in the 24 assembled master plates was sequenced, the sequences were matched against those in GenBank using the Rocket program. Only 79% of the clones matched the expected genes given in the Research Genetics database (Table 2.3). The other 21% either matched a contaminant from another clone or did not match anything in the Research Genetics database. We reported these findings in a

Rocket analysis results	Number of Clones	Percent
Identified Match	1824	79
Identified contaminant	474	20.5
Present in vendor clone library	279	• 58
Not present in vendor clone library	195	• 42
No BLAST results	6	0.5
Total number of clones analyzed	2304	100

Table 2.3 Results of data analysis by Rocket. (The symbol • indicates the proportion of contaminant identified.)

paper published in *Biotechniques* (Taylor *et al.*, 2001). In an article published in *Nature* (2001) entitled "When the chips are down," it was reported that Zacharewski and colleagues also found mistakes in the identity of up to 30% of the cDNAs in a library purchased from Research Genetics.

Many steps are involved in producing cDNA microarrays. Each step that involves manual transfer of clones (usually with a multichannel pipettor) can potentially introduce unknown errors. For instance, the frozen cultures were stabbed with a 12-channel pipettor and then swirled in the corresponding row of LB medium. Despite care, the row from the original culture plate may not have been lined up with the same row in the LB plate, and any inversions could have gone unnoticed until the quality control sequencing step. In addition, during the plate assembly step, many partial plates were converged by hand into one master plate. A further vulnerability was that database tracking was a particular challenge for these individual clones.

Thus, errors could have occurred at several places between the production of the Research Genetics clone library and the printing on the glass slides. Because database tracking for bacterial growth, PCR amplification and purification, and robotic merging into 384-well plate format does not recognize pipetting errors, contamination, or incorrect data from Research Genetics, it is essential that these mistakes be identified during the final stages of cDNA microarray production.

2.2   LONG-OLIGO ARRAYS

Although long-oligo microarrays compare well with cDNA microarrays, they also have some advantages over cDNA microarrays. First, they require no bacteria handling and clone amplification, which reduces the problems with clone contamination and practically eliminates the need for sequence verification. Second, long-oligo arrays are more specific than cDNA arrays when it comes to analyzing gene family members with $> 70\%$ sequence homology that can cross-hybridize with each other. This is because oligo sequences can be selected based on minimal homology with other genes in GenBank. Third, oligo syntheses are performed in a 96-well format and tracking can be done with a bar-coded plate, thereby reducing errors in gene identification. However, this is still not an error-free process, as the plates could mistakenly be loaded in reverse.

The oligos can be designed in-house and synthesized by commercial vendors. Alternatively, biotech companies have generated preassembled oligo libraries that consist of up to 20K unigenes (Table 2.4), providing a further incentive for microarray facilities to switch from cDNA arrays to long-oligo arrays. In addition, these oligos can be delivered in 384-well plates of investigators' choosing, thereby eliminating the need to merge 96-well plates and eliminating another potential step for error.

Vendor	Organism	Number of Oligos	Length
Sigma-Genosys	Human	18,861	60-mer
(Compugene)	Rat	7,793	65-mer
	Mouse	21,997	65-mer
	Zebra fish	16,399	65-mer
	Bacillus subtilis	4,116	65-mer
	Escherichia coli	6,048	65-mer
MWG Biotech	Human	29,952	50-mer
	Rat	9,984	50-mer
	Mouse	29,952	50-mer
	Zebra fish	14,067	50-mer
	Helicobacter pylori	1,920	50-mer
	Yeast	6,368	50-mer
	Escherichia Coli	6,336	50-mer
	Arabidopsis	19,968	50-mer
	H. hepaticus	1,921	50-mer
	Camphylobacter jejuni	1,648	50-mer
Qiagen	Human	34,580	70-mer
(Operon)	Rat	7,137	70-mer
	Mouse	31,769	70-mer
	Zebra fish	3,479	70-mer
	Yeast	6,398	70-mer
	Escherichia coli	6,002	70-mer
	Arabidopsis	26,114	70-mer
	Caenorhabdis elegans	19,873	70-mer
	Camphylobacter jejuni	1,601	70-mer
	Candida albicans	6,266	70-mer
	Drosophila	14,593	70-mer
	Human influenzae	1,714	70-mer
	Malaria	7,394	70-mer
	Pig	13,287	70-mer
	Tuberculosis	4,295	70-mer
	Salmonella	5,578	70-mer
Clontech	Human	23,000	80-mer
Illumina	Human	22,740	70-mer

Table 2.4 Partial list of commercial oligo libraries.

The method of oligo synthesis also varies with advantages and disadvantages to each. For example, Affymetrix genechips use the *in situ* synthesis of oligos on a solid support by photolithography and solid-phase chemistry to yield short [20–25 nucleotides (nt)], single-stranded oligos. A shortcoming of such short oligos is that there are more chances for unspecific hybridization to occur, and therefore many (up to 20) different short nucleotides are needed for a single gene. The poor coupling efficiency (95%) of *in situ* DNA synthesis also puts limits on the full length of the oligo, in that about 25% of oligos maintain their full length after 25 cycles of synthesis. In contrast, long oligos generated by phosphoramidite synthesis have a 99.4% coupling efficiency, and a 70-mer oligo synthesized in this way contains 65% of the full-length product. Thus, long-oligo microarrays are more likely to generate information from specific hybridization.

Long oligos can also be synthesized in various forms. They can be synthesized as unmodified oligos and then deposited onto poly-L-lysine-coated glass slides. They can also be modified with an amine group at the 5´ end, which enables the oligo to bind to aldehyde-coated glass slides (discussed further in Section 2.3).

When genome-wide arrays are desired, long-oligo microarrays may be more suitable to completely sequenced genomes, as they require sequence information. In addition, long oligos that are sequence-optimized to minimize cross-hybridization are designed to be optimally sensitive and specific to target genes. Sensitivity is gauged by the signal intensity of the particular probe after hybridization, while specificity is determined by the ratio of specific to nonspecific hybridization. Some critical characteristics that determine the sensitivity and specificity of long-oligo arrays are the length and purity of the oligos, the number of oligos deposited on the array, and the attachment efficiency.

2.2.1 Optimal length of the oligos

Several factors are considered when deciding on the optimal length of the long oligos. In general, the longer the oligos, the more efficient is the hybridization. However, one study showed that the length-dependent hybridization efficiency plateaued at 712 bases, above which hybridization efficiency decreased (Stillman and Tonkinson, 2001). At the same time, the current oligo-synthesis technologies have their limitations, including the fact that the efficiency with which completed oligos can be generated decreases with length. Most oligo synthesis companies limit the length to ~ 100 bases, when failure with complete synthesis begins to be a problem. The cost of oligo synthesis also increases with length. The oligo libraries from commercial sources range from 50 to 80 bases. Oligos of 50 mers have been examined in a number of studies, and some have compared the hybridization behaviors of oligos of 50 and 70 mers.

To gain insight into the behaviors of long-oligo microarrays, we performed a relatively comprehensive comparative study evaluating the effect of length and the numbers of oligos that are printed on hybridization efficiency. Thirty genes expressed at different levels in the target sample cell line (RKO colon cancer cell line) were chosen for this study. DNA sequences were chosen for the desired genes on the basis of their GC content (which ranged from 40% to 60%), the 3′ end biased sequence of the transcript (within 300 to 800 bases), and their least homology with other genes to minimize the possibility of cross-hybridization. Four lengths of oligos (30, 40, 50, and 70 nt) were commercially synthesized for each gene. The oligos were then purified by reverse phase cartridge purification.

A microfluidic analysis was next performed to check the quality and quantity of the oligos. Samples were assayed on a 2100 Bioanalyzer (Agilent, Foster City, California). A modified DNA 500 Assay kit utilizing the single-strand specific dye from the RNA 6000 Nano kit was used following the procedure for the RNA assay (catalog no. 5064-8229). The analysis showed that most of the oligos were of the right size and purity.

The cartridge-purified, unmodified oligos were spotted onto a poly-L-lysine-coated glass slide using the Genomic Solutions Flexys arrayer with 48 pins (Ann Arbor, Michigan). For all four lengths of oligos, the printing was done at four different oligo concentrations (20, 30, 40, and 50 μM), in array buffer containing 50% DMSO. Attachment was achieved by incubating the spotted slides at 65°C for 90 minutes or by cross-linking using ultraviolet light at 650 μJ. Total RNA from the target cell line was extracted using the RNAeasy kit (Qiagen) and reverse-transcribed to cyanine dye-labeled cDNA. The slides were then hybridized with cyanine-labeled cDNA in Expresshyb solution (Clontech Laboratories, Inc., Palo Alto, California) for 16 hours at 42°C in a humidified chamber. After hybridization, the slides were washed at 37°C once in 1× SSC (150 mM sodium chloride, 15 mM sodium citrate) plus 0.01% SDS, once in 0.2× SSC plus 0.01% SDS, and twice in 0.1× SSC sequentially for 2 minutes each.

After hybridization and washing, the slides were scanned using the Gene-Tac LSIV laser scanner (Genomic Solutions), and the signal intensities were quantified using the ArrayVision Spot finding program (Imaging Research Inc., St. Catherines, Ontario, Canada). As the array had duplicate spots for each gene, the signal intensities of the duplicate spots for samples with varying oligo lengths and varying concentrations were averaged for further analysis. The relative intensities were then calculated by comparing them with the signal intensity at the lower concentration (20 μM) or the shortest oligo length of 30 mer, both of which were set at 1.

The findings from some studies have suggested that long oligos (as long as 50 nt) give satisfactory microarray results. However, most of these studies have focused on highly expressed genes that have high signal intensities on microarrays. Because genes expressed at low levels pose more of a challenge in terms of the accuracy of microarray measurement and analysis, this impression may not be an accurate one. Indeed, some commercial vendors include only oligos that represent the most abundantly expressed genes in the testing oligo plates for the potential user's evaluation—a practice that may be misleading. Therefore, to get a more representative picture of the accuracy of the microarrays, it is important to evaluate genes that are expressed at low, medium, and high levels. Of the 30 genes in our evaluation, 8 were expressed at high and 7 at low levels; the remaining 15 showed either medium or no expression.[1] Oligos corresponding to all 30 genes were arrayed at four different concentrations (20, 30, 40, 50 μM) and four different lengths (30, 40, 50, 70 nt). A total of 480 oligo samples in duplicate were spotted onto poly-L-lysine-coated glass slides and hybridized with Cy-dye-labeled cDNAs from the RKO cell line.

We observed that both the length of the oligo and the concentration of the probes spotted on the glass appear to influence the signal intensities. Figure 2.1 shows the signal intensities of the spots yielded by the various lengths of oligos after hybridization for each gene at a 50-μM probe concentration. For those genes expressed in the RKO cells, the signal intensity yielded by a 70-nt oligo was greater than that yielded by the others. The data also suggest that this increase in signal intensities was more pronounced for the low-expressing genes. Specifically, the average increase in signal intensities for low-expressing genes (signal intensity, < 200) yielded from a 30-nt to 70-nt oligo was 13 ± 0.1 as opposed to 1.8 ± 0.2 for high-expressing genes.

2.2.2   Probe concentration

We also evaluated the effect on the signal intensities measured after hybridization of the printed number of oligo probes on the array. The general assumption was that because the probes on the slide are printed in an excess concentration compared with the amount of the target DNA in the sample, the hybridization signal intensities would be in the linear range of detection, depending mainly on the target cDNA concentration, and that this would follow a pseudo-first-order kinetics instead of second-order kinetics, in which the final product would depend on the concentration of two reactants instead of one. However, this assumption has not been extensively tested.

Figure 2.2 shows the relative fold increase in the signal intensity as a function of four different concentrations. The data from a set of low-expressing genes showed that in the concentration range used in this study, the effect on signal intensity was highest (one- to fourfold) at the highest concentration. A similar but less pronounced effect was seen for the high-expressing genes (one- to twofold). Thus, the concentration of the probe did not seem to

---

[1] We could not assess the effect of oligo length on the signal intensity for those genes that exhibited virtually no detectable signal intensity.

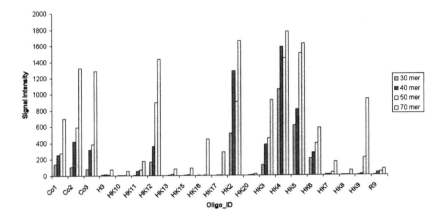

Fig. 2.1 The effect of oligo length on detected signal intensities. For each gene evaluated, the signal intensities from the four probes of different length are plotted on the bar graph. See insert for a color representation of this figure.

have a significant effect on the signal intensity, although a fourfold increase in low-expressing gene hybridization may be beneficial.

We therefore concluded that longer oligos are especially beneficial for low-expressing genes. Considering that these genes are the most challenging to detect accurately, long-oligo arrays with 70–80 mers are a better option. However, this benefit may begin to wane after that, because full-length oligos may become lost due to limitations in the efficiency of oligo synthesis.

2.2.3 Probe locations in the gene and orientation

Although, as we noted previously, oligo microarrays have several advantages over cDNA microarrays, there are several quality control issues that need to be considered in the construction of oligo microarrays. One of these issues is the selection of probe sequences in terms of their locations in the genes once the specificity and a reasonable GC content (40–60%) are satisfied. Many microarray procedures use a direct-labeling strategy starting from the poly-dT primers in a reverse transcription (see Chapter 3). However, because the reverse transcription reaction is not efficient when cyanine dye-labeled nucleotides are used, the labeled cDNA products tend to be short. Therefore, most oligo probes for oligo microarrays are selected from the 3´ end of the genes, normally within the 1-kb region of the 3´-most end from the poly-A tails to guarantee that the cyanine-labeled target sequences are long enough to include the probe sequence. With other procedures, such as indirect labeling, this requirement is less of an issue. New compounds, such as fluorescent monofunctional cisplatin derivatives, have now been developed to label nu-

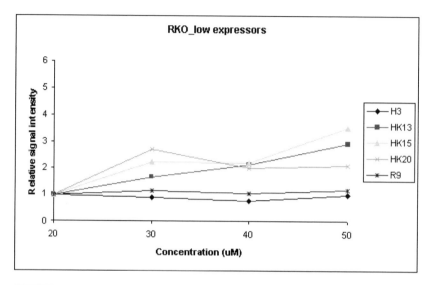

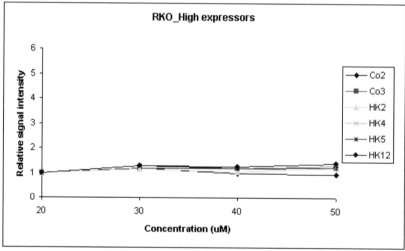

Fig. 2.2   The effect of probe amount on detected signal intensities. Relative intensities for 20, 30, 40, and 50 μM concentration of the probe in comparison to 20 μM. See insert for a color representation of this figure.

cleotides for microarrays. Hagedoorn *et al.* (2003) found that this labeling methodology eliminated the problem of the 3´-end bias that is associated with the conventional enzymatic oligo-dT-primed reverse transcription of mRNAs. In addition, Gupta *et al.* (2003) suggested that this type of direct labeling of mRNA yields highly precise and unbiased expression data. Therefore, it is important to know the hybridization procedures that have been used when selecting oligos for microarray production. At present, however, because oligos at the 3´ end are suitable for most protocols, they are still the best choice.

Another important issue is the orientation of the oligos—sense or antisense. Most oligo libraries use a sense orientation, but some studies require antisense oligos for a specific purpose. For example, Lee *et al.* (2001) used an antisense *Bacillus subtilis* genome array to study gene expression and discovered a novel transcript within the intergenic regions and on the opposite coding strand of known open reading frames.

Mistakes can be costly if the sense and antisense orientations are mixed up. For example, as noted in an article in *Nature,* "When the chips are down" (2001), Affymetrix in early 2001 had to announce that a third of the sequences on a set of mouse arrays were from the wrong strand from a public sequence database, and thus did not detect the target transcripts. The problem has since been corrected. However, similar mistakes may still occur, especially in microarrays designed for species for which there are incomplete and poorly annotated sequence databases.

2.3 SLIDE COATING

Many microarray facilities, including ours, print either cDNA or long-oligos on poly-L-lysine-coated glass slides, which have an amine surface. These coated slides can be purchased from commercial sources or produced in-house. An amine surface is positively charged and binds to the negatively charged phosphate groups on DNA molecules. The process of coating slides with poly-L-lysine is simple and inexpensive. The DNA is simply attached to the glass by placing the slide in a drying oven and exposing it to ultraviolet light, which causes cross-linking (simple cross-linking instruments can be purchased). Although blocking can be done after the microarrays are printed, because most hybridization buffers contain SDS, the sulfate group of which provides a negative charge that prevents the negatively charged phosphate backbone of DNA from binding to the slide surface, a separate blocking step is generally not necessary to reduce background fluorescence signals significantly. Our core laboratory has been coating slides using a protocol adapted from the one developed by DeRisi in Brown's laboratory at Stanford University. The performance of our slides is comparable to that of most commercial slides but cost approximately ten times less. Because the amine group is hydrophilic, the printed spots on these slides tend to be bigger due to diffusion, which in turn may reduce the maximum spotting density.

The second most popular coating is aldehyde. For DNA to bind to aldehyde groups on the slides, a covalent bond must form between the amine group on the DNA and the aldehyde group on the slide. Many commercial oligo libraries have a 5´ amine group on the oligos that can be printed onto the aldehyde-coated slides. One advantage of this is that only one end of the probe is then bound to the glass surface and the rest of the DNA molecule is in a more free form, which may increase its accessibility to the targets for hybridization. However, because aldehyde-coated microarray slides have an elevated level of background signals, the slides need to be exposed to bovine serum albumin (BSA) to reduce the background fluorescent signals. BSA is used because it contains lysine residues that bind to the reactive aldehyde group, which prevents the binding caused by the amine group present in DNA. Another advantage of aldehyde-coated slides is that they are hydrophobic and thus have relatively smaller spot diameters, which allows greater printing densities than those allowed by amine-coated slides.

Yet another coating is epoxy, which can effectively bind to both DNA and proteins. Some groups choose to coat their slides with both epoxy and amine so that they can observe the synergistic effects of the two on DNA immobilization and hybridization (Chiu et al., 2003). Slides with different surfaces are available from commercial sources, with prices ranging from $2 to $20 apiece (Table 2.5).

Microarrays can also be constructed without modification of the glass surface. For example, an active silyl moiety can be covalently linked to the cDNA or oligo to produce silanized DNAs, which can be immobilized on unmodified glass slide surfaces immediately after deposition by microarray spotters (Kumar et al., 2000).

2.4   SLIDE AUTOFLUORESCENCE

Most glass-based microarrays use fluorescence to gauge the amount of the transcript binding to the probes on the array. Therefore, it is important that only the hybridized transcripts fluoresce. Unfortunately, many substances autofluoresce after laser excitation, which would interfere with the results. Because these substances are present on a fraction of commercial glass slides, all glass slides should be washed with diluted acid before coating. After the slides are coated with 0.1% poly-L-lysine, we recommend that they be scanned with a laser scanner to weed out slides that autofluoresce. Figure 2.3 shows some examples of such slides. Although the fraction of slides that autofluoresce differs from batch to batch, the average is about 20%, and using these slides for microarrays would diminish the quality of downstream analysis. Considering the low cost of regular glass slides and the relatively higher price of printed microarray slides, prescreening the slides is a prudent step.

Another problem is the more uniform autofluorescence that can be emitted by coating material or result from the binding of labeled material to the coat-

Manufacture	Slide name	Active group in coating	Designed for:	Cost/slide ($)
Corning	GAPSII	Amine	cDNA	11
	UltraGAPS	Amine	cDNA	13
			Oligo	
Telechem	Super Amine	Amine	cDNA	8.4
			Oligo	
	Super Aldehyde	Aldehyde	Amino-mod. cDNA and oligo	9.4
			Proteins	
	Super Epoxy	Epoxy	Oligo	9
			Amino-mod. oligo	
			cDNA	
	Mirrored Amine	Amine	cDNA	29
			Oligo	
	Mirrored Aldehyde	Aldehyde	Amino-mod. cDNA	30
			Amino-mod. oligo	
			Proteins	
	Super Gold	Au layer	cDNA	42.4
			Proteins	
			Cell culture	
FullMoon Biosciences	FMB cDNA	Amine	cDNA	10
	FMB oligo	Epoxy	Oligos	11
			Amino-mod. oligo	
	Power Matrix-amino	Matrix	Amino-mod. oligo	13
	Power Matrix	Matrix	Oligo	13
	protein	Matrix	Protein lysates	13
Schleicher & Schuell	FAST	Nitrocellulose	Protein	12.5

Table 2.5 Commercially available coated slides.

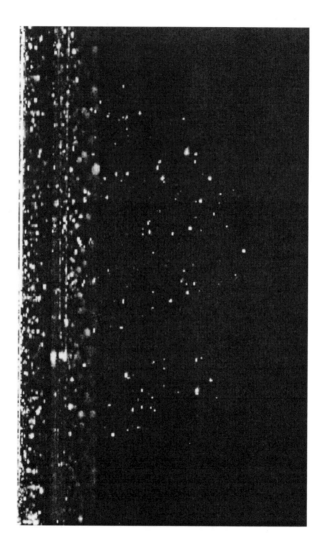

Fig. 2.3 Detection of autofluorescence by prescanning slides. An example showing a small region of a slide with high autofluorescence.

ing material. This is especially likely when aldehyde coating material is used. There are several ways to reduce this background autofluorescence. SDS, which is present in most hybridization buffers, is often sufficient to reduce the autofluorescence that arises from amine-coated slides. BSA is very effective in reducing the background fluorescence that comes from aldehyde-coated slides. Sodium borohydride can also be used to reduce autofluorescence on microarrays (Raghavachari *et al.*, 2003). Another common practice is to preheat the microarray slides to 65°C immediately before adding the hybridization mixture. This prevents the labeled targets from binding to the reactive groups on the slides but allows the blocking agent (such as SDS) to bind, thus reducing the background signals.

Some specific slides show other types of background signals. For example, Martinez *et al.* (2003) identified a spot-localized, contaminating fluorescence in the Cy3 channel on several commercial and in-house-printed microarray slides. By doing mock hybridizations without the labeled target, they determined that prehybridization scans could not be used to predict the contribution of this contaminating fluorescence after hybridization because the change in spot-to-spot fluorescence after hybridization was too variable. Nonetheless, they found that using Corning UltraGAPS slides (Corning, New York) and exposing the slides to air for 4 hours prior to printing significantly reduced contaminating fluorescence intensities to approximately that of the surrounding glass.

2.5 PRINTING QUALITY CONTROL

Once high-fidelity DNA (cDNA or long oligos) is available and cleanly coated slides are ready, microarrays are ready to be printed with a robotic printer or spotter. There are a number of different printers on the market and there are different issues specific to each that will not be covered here, but that need to be considered. Detailed information on these issues can be found in other sources, such as the book by Schena (2003). We discuss here only the quality control issues common to the use of all printers.

2.5.1 Arraying buffer

First, the cDNA or long oligos need to be dissolved in a suitable buffer for printing. There are two major types of buffers that have been used. One is an SSC-based salt buffer, which is commonly used for quill-pin printers. The other is a DMSO-based buffer. The DMSO drastically slows evaporation, making it particularly useful for solid-pin printers, as the latter typically require more time for printing than quill-pin printers. However, a DMSO-based buffer is not suitable for quill-pin printers. Some researchers use a detergent such as sarkozyl to increase the surface tension for a ring-and-pin spotter. It is important to prevent buffer evaporation during the lengthy time it can take to print high-density arrays. The humidity of the printing chamber should therefore be controlled. This may not be a major issue in cities such as Houston in the summer, when humidity in the laboratory can reach 55%. In cities with lower humidity, however, a humidifier is needed to prevent the buffer from evaporating.

2.5.2   Printing spot quality and dropout problems

Several problems may occur when the arrayer prints thousands of probes onto the coated glass slides. Occasionally, the printed spot morphology can be very poor if some of the printing pins are defective. This is a particular problem for quill pins, which may become clogged and fail to print, resulting in so-called dropouts on the arrays. Figure 2.4 shows a magnified image of a defective quill pin. Note that the bent tip will prevent takeup of the arraying buffer, producing spot dropouts. Some researchers have had problems with printing failures that appear to result from surface tension problems with the printing buffer. This can happen when the quill pins do not take up enough DMSO-based arraying buffer. Some have experienced similar problems with phosphate-based arraying buffer. However, this is not a problem with an SSC-based arraying buffer. Some pins can also be very sensitive to the volume of arraying buffer in the wells of plates. There has to be a large volume in the wells or the pins may not get filled evenly. Thus, there has to be a large amount of DNA for the printing process. Complete consumption of some probes in the 384-well DNA source plate can also lead to selected dropouts. These spots will not produce data regardless of the expression status of these genes. Another possible problem is that some printed DNAs may not stick well to the glass, and thus get washed away during the hybridization step.

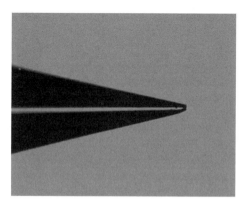

Fig. 2.4 A magnified image of a defective quill pin, which would result in spot dropouts.

Although there exist some algorithms that have been proposed to estimate the missing data resulting from dropouts (see Chapter 6), dropouts often lead to elimination of these genes from the entire analysis. Thus, it is important to evaluate the printing process, which can be done in several ways.

A simple way is to breathe onto the slide to form a thin layer of moisture; spots where DNAs are printed appear as clear spots. However, this is a very crude method and does not show the quality of the spots and the extent of missing spots unless a large number of spots are missing. Another way is to look at the printed slides under a microscope. The salt in the printing leaves a mark on the slide which can be seen under a microscope. However, this cannot show the amount of DNA that has been printed or how tightly the DNA is bound to the glass surface.

A more reliable way to evaluate the quality of microarray printing is to stain the slides with a fluorescent dye. The most commonly used dye is SybrGold (Molecular Probes). Another commonly used dye is SYBR green II. After each batch of microarrays is printed, one slide from the batch can be stained with the dye. After washing, the slide is then scanned in a laser scanner so that the spot morphology and the amount of DNA printed on the spots can be evaluated. An example of such a slide is shown in Figure 2.5. Information on the dropouts can then be stored for incorporation into downstream data analysis. Measures can also be taken after this step to identify the sources of the dropouts and the bad spot morphology (e.g., a pin needs to be cleaned or replaced or some wells need to be refilled with DNAs).

Although SybrGold staining is relatively noninvasive and the slides can be reused again, it can tell nothing about the tightness of the binding of DNAs to the slides and how accessible the DNAs are for hybridization. Our group and others have used another simple method to check this. First, since all cDNAs on our microarrays are PCR-amplified by the same set of primers (except for β-actin and GAPDH, which are generated by their specific primers), all corresponding cDNAs printed on our microarray slides should hybridize to the primer labeled with either cyanine-3 or cyanine-5. This primer can then be used to check the quality of printing and, especially, the attachment of cDNA probes on the array after hybridization. In our microarray layout (Figure 2.6), the primer target can also be used to check the specificity of hybridization, because it should not hybridize to cDNA spots that are not generated by the common primers.

These tests can confirm that the cDNAs are effectively deposited on the array and that they have remained attached after all processing and hybridization procedures. We found that cyanine-dye-labeled primer target hybridizes to most of the spots on the array, but the signal intensities show some variability (Figure 2.6), although this might also be caused by adjacent sequences on the different probes on the array.

To evaluate the slide-to-slide variability of spots, we carried out 14 hybridizations on 14 slides using labeled primer targets and calculated the coef-

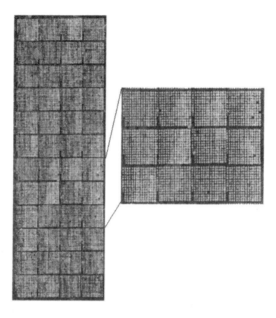

Fig. 2.5 An example of a printed slide after staining with SybrGold.

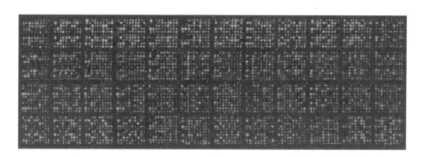

Fig. 2.6 Cy3-labeled primer hybridized to every spot generated by PCR using the same primer.

ficient of variation (CV) for all the 4608 spots on the array. The CVs ranged from 0 to 20.2%, with a median CV of 6.5% and a mean CV of 6.8%.

Despite close monitoring of the printing process, it is difficult to completely prevent dropouts from occurring. Recently, Khan *et al.* (2003) proposed multiplexing more than one gene into each spot to efficiently provide robustness to dropouts in microarray production. In this way, the expression information for each gene could then be decoded via mathematical methods. In addition to this, Shmulevich *et al.* (2003) showed that construction of this type of composite microarray allowed multiple measurements of each gene, which results in greater overall accuracy (see Chapter 6).

2.6 SUMMARY

In summary, microarray production is one of the most important steps in microarray experiments, with quality control being an important component of the process. There are now many biotechnology companies that have produced competing products, but because the technology is new, microarrays are an expensive technology and few researchers can carefully evaluate all of the different platforms. Nevertheless, information has been accumulated during the past several years and a number of reliable quality control procedures have been developed. In particular, for cDNA arrays, it is important to verify the correct identities of the clones printed on the array, if their names are given to researchers. It is also advisable to use a robotic liquid-handling system to handle high-throughput liquid transfer. In terms of oligo microarrays, it is important to know the effect of oligo length on the detection of signals important for particular experiments. It is also worthwhile to monitor printing to prevent dropout problems.

REFERENCES

1. Alon U, Barkai N, Notterman DA, Gish K, Ybarra S, Mack D, Levine AJ. (1999) Broad patterns of gene expression revealed by clustering analysis of tumor and normal colon tissues probed by oligonucleotide arrays. *Proc Natl Acad Sci USA* 96: 6745–6750.

2. Altschul, SF, Gish W, Miller W, Myers EW, Lipman DJ. (1990) Basic local alignment search tool. *J Mol Biol* 215:403–410.

3. Battaglia C, Salani G, Consolandi C, Bernardi LR, De Bellis G. (2000) Analysis of DNA microarrays by non-destructive fluorescent staining using SYBR green II. *Biotechniques* 29:78–81.

4. Benson DA, Karsch-Mizrachi I, Lipman DJ, Ostell J, Rapp BA, Wheeler DL. (2000). GenBank. *Nucleic Acids Res* 28:15–18.

5. Bittner M, Meltzer P, Chen Y, Jian Y, Seftor E, Hendix M, Radmacher M, Simon R, Yakhini Z, Ben-Dor A, Sampas N, Dougherty E, Wang E, Marincola F, Gooden C, Lueders J, Glatfelter A, Pollock P, Carpten J, Gillanders E, Leja D, Dietrich K, Beaudry C, Berens M, Alberts D, Sondak V, Hayward N, Trent J. (2000). Molecular classification of cutaneous malignant melanoma by gene expression profiling. *Nature.* 406:536–540.

6. Chiu SK, Hsu M, Ku WC, Tu CY, Tseng YT, Lau WK, Yan RY, Ma JT, Tzeng CM. (2003) Synergistic effects of epoxy and amine-silanes on microarray DNA immobilization and hybridization. *Biochem J* 374:625-32.

7. Clark EA, Golub TR, Lander ES, Hynes RO. (2000) Genomic analysis of metastasis reveals an essential role for RhoC. *Nature* 406:532–535.

8. Diehl F, Beckmann B, Kellner N, Hauser NC, Diehl S, Hoheisel JD. (2002) Manufacturing DNA microarrays from unpurified PCR products. *Nucleic Acids Res* 30:e79.

9. Fuller GN, Rhee CH, Hess KR, Caskey LS, Wang R, Bruner JM, Yung WKA, Zhang W. (1999) Reactivation of Insulin-like Growth Factor Binding Protein 2 Expression in Glioblastoma Multiforme: A Revelation by Parallel Gene Expression Profiling. *Cancer Res* 59:4228–4232.

10. Gupta V, Cherkassky A, Chatis P, Joseph R, Johnson AL, Broadbent J, Erickson T, DiMeo J. (2003) Directly labeled mRNA produces highly precise and unbiased differential gene expression data. *Nucleic Acids Res* 31:e13.

11. Hagedoorn R, Joseph R, Kasanmoentalib S, Eilers P, Killian J, Raap AK. (2003) Chemical RNA labeling without 3´ end bias using fluorescent cis-platin compounds. *Biotechniques* 34:974–6, 978, 980.

12. Halgren RG, Fielden MR, Fong CJ, Zacharewski TR. (2001) Assessment of clone identify and sequence difelity for 1189 IMAGE cDNA clones. *Nucleic Acids Res* 29:282–288.

13. Hughes TR, Mao M, Jones AR, Burchard J, Marton M J, Shannon KW, Leftowitz S M, Ziman M, Schelter JM, Meyer MR, *et al.* (2001) Expression profiling using microarrays fabricated by an ink-jet oligonucleotide synthesizer. *Nature Biotechnol* 19:342–347.

14. Jobs M, Fredriksson S, Brookes AJ, Landegren U. (2002) Effect of oligonucleotide Truncation on Single-Nucleotide Distinction by Solid-Phase Hybridizaion. *Anal Chem* 74:199–202.

15. Kane MD, Jatkoe TA, Stumpf CR, Lu J, Thomas JD, Madore SJ. (2000) Assessment of the sensitivity and specificity of oligonucleotide (50mer) microarrays. *Nucleic Acids Res* 28:4552–4557.

16. Khan AH, Ossadtchi A, Leahy RM, Smith DJ. (2003) Error-correcting microarray design. *Genomics* 81:157–65.

17. Kumar A, Larsson O, Parodi D, Liang Z. (2000) Silanized nucleic acids: a general platform for DNA immobilization. *Nucleic Acids Res* 28:e71.

18. Lee JM, Zhang S, Saha S, Santa Anna S, Jiang C, Perkins J. (2001) RNA expression analysis using an antisense Bacillus subtilis genome array. *J Bacteriol* 183:7371–7380.

19. Lennon GG, Auffray C, Polymeropoulos M, Soares MB. (1996) The I.M.A.G.E. Consortium: An Integrated Molecular Analysis of Genomes and their Expression. *Genomics* 33:151–152.

20. Lipshultz RJ, Fodor SP, Gingeras TR, and Lockhart DJ. (1999) High density synthetic oligonucleotide arrays. *Nature Genet* 21:20–24.

21. Lockhart DJ, Barlow C. (2001) Expressing what's on your mind: DNA arrays and the brain. *Nature Rev Neurosci* 2:63–68.

22. Madden TL, Tatusov RL, Zhang J. (1996) Applications of network BLAST Server. *Methods Enzymol* 266:131–141.

23. Martinez MJ, Aragon AD, Rodriguez AL, Weber JM, Timlin JA, Sinclair MB, Haaland DM, Werner-Washburne M. (2003) Identification and removal of contaminating fluorescence from commercial and in-house printed DNA microarrays. *Nucleic Acids Res* 31:e18.

24. News feature: When the chips are down. *Nature* 410:860–861, 2001.

25. Peterson AW, Heaton RJ, Georgiadis RM. (2001) The effect of surface probe density on DNA hybridization. *Nucleic Acids Res* 29:5163–5168.

26. Raghavachari N, Bao YP, Li G, Xie X, Muller UR. (2003) Reduction of autofluorescence on DNA microarrays and slide surfaces by treatment with sodium borohydride. *Anal Biochem* 312:101–5.

27. Ramdas L, Coombes KR, Baggerly K, Hess K, Abrruszzo L, Zhang W. (2001).Sources of nonlinearity in cDNA microarray expression measurments. *Genome Biol* 2:47.1–47.7.

28. Relogio A, Schwager C, Richter A, Ansorge W, Valcarcel, J. (2002) Optimization of Oligonucleotide-based DNA microarrays. *Nucleic Acids Res* 30:e51.

29. Rouillard J-M, Herbert JH, Zuker M, (2002). OligoArray: genome-scale oligonucleotide design for microarrays. *Bioinformatics* 18:486–487.

30. Schena M. (2003) *Microarray analysis.* John Wiley & Sons, New York.

31. Schena M, Shalon D, Davis RW, Brown PO. (1995). Quantitative monitoring of gene expression patterns with a complementary DNA microarray. *Science* 270:467–470.

32. Schuler G.D. (1997). Pieces of the puzzle: expressed sequence tags and the catalog of human genes. *J Mol Med* 75:694–698.

33. Shmulevich I, Astola J, Cogdell D, Hamilton SR, Zhang W. (2003) Data extraction from composite oligonucleotide microarrays. *Nucleic Acids Res* 31:e36.

34. Shoemaker DD, Schadt EE, Armour CD, He YD, Garrett-Engele P, McDonagu PD, Loerch PM, Leonardson A, Lum PY, Cavet G, *et al.* (2001) Experimental annotation of the human genome using microarray technology. *Nature* 409:922 – 927.

35. Southern E, Mir K, Shchepinov M. (1999) Molecular interactions on microarray. *Nat Genet* 21: 5–9.

36. Stillman BA, Tonkinson JL. (2001) Expression microarray hybridization kinetics depend on length of the immobilized DNA but are independent of immobilization substrate. *Anal Biochem* 295:149–57.

37. Tatusova TA, Thomas ML. (1999) Blast 2 sequences—a new tool for comparing protein and nucleotide sequences. *FEMS Microbiol Lett* 174:247–250.

38. Taylor E, Cogdell D, Coombes K, Hu L, Ramdas L, Tabor A, Hamilton S, Zhang W. (2001) Sequence verification as quality-control step for production of cDNA microarrays. *Biotechniques* 31:62–65

39. Troyanskaya O, Cantor M, Sherlock G, Brown P, Hastie T, Tibshirani R, Botstein D, Altman RB. (2001) Missing value estimation methods for DNA microarrays. *Bioinformatics* 17:520–5.

3

Quality of Microarray Hybridization

With high-quality biological material and a set of high-quality microarrays in hand, one is ready to proceed to the next step in gene expression profiling—hybridization. Hybridization is the most tedious step in microarray experiments, demanding the most patience and greatest technical skills.

The basic principle of hybridization in microarrays is the same as that of conventional Southern or Northern blotting assays. However, a key difference is that the cellular mRNAs are in the liquid phase and the probes are immobilized, and for this reason, microarrays are sometimes called *reversed-phase Southern or Northern blotting assays.* In addition, glass is more frequently used than nitrocellulose membranes as the printing substrate because it allows the high-density printing of probes within a small area. In these glass-based microarrays, the targets are labeled with fluorescent dye in the form of cyanine-3 or cyanine-5 (Cy3 and Cy5) dNTP or NTP rather than with radioactive dNTP or NTP to generate "hot" targets. An advantage of fluorescent dye-labeled nucleotides over radioactive isotope-labeled nucleotides is that they decrease the exposure of researchers to biological hazards. A further advantage is that two different fluorescent dyes, such as Cy3 and Cy5, can be used simultaneously, which allows two different samples to be directly compared on a single microarray. A drawback to the fluorescence-based method is that it is not as sensitive as the radioisotope-based method in detecting gene expression, making sensitivity a major issue in glass-based microarray experiments, especially when the biological material is limited.

3.1   RNA AND cDNA LABELING QUALITY

In Chapter 1 we touched on the quality of the biological materials to be used for microarray analyses. Not only is the quality of the RNA determined by the tissue acquisition and storage procedures, it in turn determines the quality of the data. There are several factors that affect the quality of RNA. First, RNAs are not stable. In addition, RNases in the cells are abundant and resistant to denaturation even at the boiling-point temperature. In the past, for example, researchers had to take extensive measures to clean all glassware that came in contact with RNA. However, with the advent in recent years of RNase inhibiting reagents and RNA isolation kits, it has become easier to handle RNA.

There are several well-known ways to evaluate RNA quality. The most classic one is running the RNA on an agarose gel. High-quality total RNAs should show two distinct bands, representing 28S and 18S ribosomal RNA. After ethidium bromide staining, however, the intensity of 28S rRNA should be approximately twice that of 18S rRNA, and there should be no significant amount of smaller-molecular-weight RNA species. A drawback of this method, however, is that it requires the use of a fairly large amount of the RNA sample, which in many cases cannot be spared.

With the development of a bioanalyzer system by Agilent (Foster City, California), which can analyze as little as 5 ng of RNA (for the Nanochip) or 200 pg of RNA (for the Picochip), this problem has been overcome. Needless to say, this system has now become quite popular among molecular biology researchers. Detailed information regarding how the system works can be found at the Agilent Web site (http://www.chem.agilent.com). When an RNA sample is analyzed by the bioanalyzer, one sees two peaks, one representing the 28S and the other the 18S rRNA (Figure 3.1). When the quality of the RNA is good, the ratios between the two peaks range from 1.5 to 2.0. The Agilent bioanalyzer, however, does not measure the purity of RNA samples, which needs to be done with a UV spectrometer. Good-quality RNA should have an absorbance ratio at 260 nm/280 nm above 1.8. The ratio is less if the RNA is contaminated with proteins. UV spectrometry, however, cannot assess the integrity of the RNA samples. In other words, although degraded, RNAs can be shown by UV spectrometry to be pure. Sometimes, pure RNAs do show a reading lower than 1.8; incomplete removal of the phenol from the samples and the use of DEPC-treated water can be the cause. In general, RNA dissolved in neutral TE buffer gives the most consistent reading of the 260 nm/280 nm ratio. Thus, we recommend passing isolated RNA through a column to remove all traces of phenols and other contaminating chemicals before it is evaluated.

Phenol must also be removed to obtain good labeling at the next step of microarray experiments. Perhaps because of the poor labeling efficiency of cyanine-dye nucleotides, the labeling reaction is also highly sensitive to impurities in the RNA solution. Phenol is often the culprit. There have also

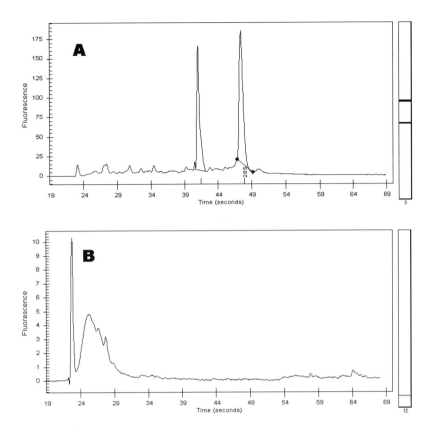

Fig. 3.1 2100 Bioanalyzer data. A: Good RNA from snap-frozen preparation of a clinical sample with 28S/18S = 1.24; B: RNA from another preparation with 28S/18S = 0, indicating severe RNA degradation.

been cases of investigators dissolving RNA in an SDS-based buffer, which is clearly not appropriate for an enzymatic reaction. Therefore, the appropriate buffer for RNA is also important to the labeling reaction. TE and water both work well.

3.2   SPECIFICITY VERSUS SENSITIVITY

There are many considerations concerning the use of glass-based microarrays. For example, many conditions have been tried to increase the sensitivity of the microarrays and thereby detect sufficient signals. On the other hand, sensitivity could be increased at the expense of specificity. Without specificity, sensitivity has little meaning. Therefore, some simple quality control measures should be taken to prevent nonspecific hybridization.

We designed a simple method to evaluate hybridization specificity and along with that, microarray printing and cDNA attachment after hybridization (Chapter 2, Figure 2.6). In particular, we used one primer both to amplify all cDNA clones on the array [except β-actin (ACTB) and GAPDH] and to hybridize the microarray slides. We labeled β-actin with Cy5 as a specific target of hybridization, and thus hybridization specificity was shown by the visualization of fluorescent signals solely from the spots where the corresponding probes were printed. In particular, the β-actin target was generated from K562 leukemia cells by RT-PCR, during which Cy5 dCTP was incorporated. β-actin was labeled with Cy5 and hybridized to the microarray slides. β-actin cDNA was deposited at the upper left corner of eight grids. There were also two spots of γ-actin probe on the array, which share 82% sequence homology with the β-actin gene. This allowed us to evaluate the extent of cross-hybridization with genes from the same gene family. Cy5-labeled β-actin and Cy3-labeled common primers were then cohybridized to a microarray slide. After the slide was washed and scanned, the hybridization results were examined.

As shown in Figure 3.2, the Cy5-labeled β-actin target hybridized strongly with the eight corresponding β-actin cDNAs and weakly with the two γ-actin cDNAs on the array, whereas the Cy3-labeled primer target hybridized with all corresponding spots on the array but not with the β-actin and GAPDH cDNA. To determine the extent of cross-hybridization, we quantified the intensities of the fluorescent signals from the β-actin and γ-actin spots. The average cross-hybridization to the two γ-actin probes generated an intensity that was 25% of that of the eight β-actin probes on the array.

The simplicity of this method is shown by the fact that if mistakes had occurred, for example, the β-actin clones had been mislocated, the pattern of the hybridization signals would have been wrong. The results also showed that our hybridization condition was very specific and that gene family members shared as much as 82% sequence homology with only 25% cross-hybridization in terms of intensity.

Based on our experience, we recommend that a primer such as the one described be used to monitor probe printing and attachment on the array if cDNA probes on the array are PCR-amplified by the same set of primers and that some specific targets be used to confirm the specificity of hybridization and the correct clone arrangement on the microarrays.

3.3 AMPLIFICATION STRATEGIES

The quality and quantity of biological materials available for cDNA microarray studies is often a limiting factor. This is especially true for clinical samples, especially those obtained by fine-needle aspiration (FNA). We have ear-

A   B   C

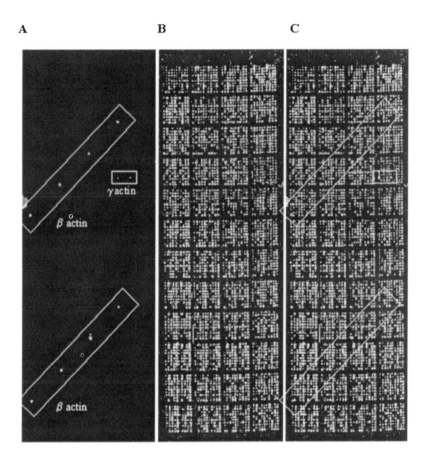

Fig. 3.2 Specific hybridization of β-actin target on the array. A β-actin target was generated by RT-PCR from K562 leukemia cell RNA. Cy5-dCTP was incorporated during PCR. The downstream primer used to generate all cDNA on the array, except for GAPDH and β-actin, was end-labeled with Cy3. Cy5-labeled β-actin and Cy3-labeled primer were cohybridized to the microarray slides. The array contains 2303 known human genes printed in replicate, in addition to positive and negative controls. After washing and scanning of the slide, the hybridization result showed that (A) the β-actin target was specifically hybridized to the corresponding spots on the array and the (B) Cy3-labeled primer hybridized to every spot generated by PCR using the same primer, indicated by green spots on the array. (C) The composite image of both Cy3 and Cy5 channels. Hybridization to β-actin and γ-actin was indicated. This figure is reproduced from Hu *et al.* (2002a) with permission from *Biotechniques*. See insert for a color representation of this figure.

lier shown that FNA samples can routinely yield good-quality RNA, but the amount of RNA obtained from FNA samples is often as low as 2 μg. A total of at least 50 μg of total RNA is required for most microarray protocols without amplification. Because of the difficulty in coming up with this much RNA, protocols have been developed to amplify RNAs isolated from limited amounts of tissues or cells. *In vitro* transcription is the most widely used of the RNA amplification protocols available. There are two means of generating a double-stranded cDNA template for the RNA polymerase: (1) a combination of reverse transcription (RT) with conventional second-strand cDNA synthesis and (2) a combination of the switch mechanism at the 5´ end of RNA templates (SMART) with RT, followed by PCR. Both protocols combine cDNA synthesis with a template-directed *in vitro* transcription reaction. During the reaction, a synthetic oligonucleotide containing a bacterial RNA polymerase promoter sequence, such as the T7 or SP6 RNA polymerase promoter sequence, is incorporated into the cDNA molecules. This method, commonly known as the *Eberwine procedure*, was developed by Eberwine and colleagues (Van Gelder *et al.*, 1990). The second strand of cDNA, which serves as the template for the RNA polymerase, can be generated either by conventional second-strand cDNA synthesis or by combining the SMART followed by PCR. Using either method, RNA is amplified after the initial step with *in vitro* transcription using RNA polymerase, and the amplified RNA (aRNA) is labeled with Cy3-dTCP or Cy5-dCTP by RT.

3.3.1   Study of amplification protocols

One possible caveat is that different amplification protocols may generate different results. This was evaluated in a study we performed in which we analyzed a set of six microarray experiments involving the use of a "regular" (unamplified) microarray experimental protocol and two different RNA amplification protocols. On the basis of their ability to identify differentially expressed genes and the assumption that the results from the regular protocol were a reliable benchmark, our analyses showed that both amplification protocols achieved reproducible and reliable results. Our results also showed that more amplified RNA could be obtained using conventional second-strand cDNA synthesis than using the combination of SMART and PCR. Therefore, when the critical issue is the amount of starting RNA, we recommend using the conventional second-strand cDNA synthesis as the amplification method. Following are some of the methods that we used in this experiment. We describe the methods in a somewhat detailed manner in order to enable the investigator wishing to compare and assess various amplification protocols to evaluate their performance properly.

3.3.1.1 cDNA synthesis   We carried out first-strand cDNA synthesis by RT in a 20-μL solution of 1 μg of oligoT$_{25-}$T$_7$ (5´-AAACGACGGCCAGTGAATTGTAATACGACTCACTATAGGGCGATT-

3´), 1 μg of total RNA, 4 μL of first-strand reaction buffer (GIBCO BRL, Invitrogen Corp., Carlsbad, California), 2 μL of 10 mM dithiothreitol (DTT; GIBCO BRL, Invitrogen), 1.0 μL of 10 mM dNTPs, 1.0 μL of SUPERase-in (Ambion, Austin, Texas), and 200 units of Superscript II reverse transcriptase (GIBCO BRL, Invitrogen).

For the conventional second-strand cDNA synthesis protocol, we added the following reagents to the 20-μL RT reaction solution: 91 μL of nuclease-free water, 30 μL of 5× second-strand buffer (GIBCO BRL, Invitrogen), 10 units of *E. coli* DNA ligase (New England Biolabs, Beverly, Massachusetts), 40 units of *E. coli* DNA polymerase I (New England Biolabs), and 2 units of RNase H (GIBCO BRL, Invitrogen). The reaction was carried out in the final volume of 150 μL at 16°C for 2 hours.

For the template-switching protocol, we included 1 μg of template-switching primer (primer sequence: 5´-AAGCAGTGG-TAACAACGCAGGGACCGGG-3´) during synthesis of the first strand of cDNA. The reaction was allowed to continue for 2 hours at 42°C. To synthesize the second strand of cDNA, we added 1 unit of RNase H (Roche, Branchburg, New Jersey) to the 20-μL RT reaction solution, and then incubated the resultant solution at 37°C for 15 minutes. We then added the following reagents: 57 μL of nuclease-free water, 10 μL of 10× PCR buffer (Roche), 10 μL of 25 mM MgCl$_2$, 1.0 μL of 10 mM dNTPs, and 5 units of Ampli-Taq Gold DNA polymerase (Roche). The reaction was allowed to continue at 95°C for 10 minutes and then for three cycles at 95°C for 1 minute, 65°C for 6 minutes, and up to 12 minutes in the final elongation cycle. We then purified the cDNA products generated from both protocols using the QIAquick PCT Purification Kit (Qiagen, Valencia, California).

3.3.1.2 *RNA amplification and target labeling*

We amplified antisense RNA by T7 *in vitro* transcription using the reagents from the MEGAscript T7 Kit (Ambion, Austin, Texas). The reaction was carried out in a total volume of 40 μL, which included 7.5 mM NTPs, 4.0 μL of 10× buffer, 4 μL of the enzyme mixture, and all the cDNA products from the cDNA synthesis. After RNA amplification, we removed the cDNA template by incubating the reaction mixture with 4 units of RNase-free DNase I (Ambion) at 37°C for 15 minutes, after which we purified the aRNA using the RNeasy Mini Kit (Qiagen). We labeled the purified 5 μg of aRNA with Cy3 or Cy5 by RT in a solution containing 2 μg of random hexamer; 4 μL of first-strand reaction buffer (GIBCO BRL, Invitrogen); 2 μL of 10 mM dithiothreitol (DTT; GIBCO BRL, Invitrogen); 1 μL of 2 mM dATP, dGTP, and dTTP and 1 mM dCTP, 1 μL of SUPERase-in^{TM} (Ambion); 1.0 μL of Cy3-dCTP or Cy5-dCTP (Amersham Pharmacia Biotech, Piscataway, New Jersey) and 200 units of Superscript II reverse transcriptase (GIBCO BRL, Invitrogen). Labeling was carried out at 42°C for 2 hours. We then purified the labeled cDNA using MicroSpin G-50 columns (Amersham Pharmacia); and reduced the volume to about 10 μL using the Speed-Vac System AES2010 (Savant Instruments, Inc. Holbrook, New York) before hybridization.

3.3.1.3 Imaging quantification   We used ArrayVision (Imaging Research, Inc., St. Catherines, Ontario, Canada) to quantify the microarray images. The signal-to-noise (S/N) ratio was calculated by dividing the background-corrected intensity by the standard deviation (SD) of the background pixels. Quantification files were loaded into S-Plus 2000 (Insightful Corp., Seattle, Washington) for data processing and analysis.

3.3.2   Results of study of amplification protocols

3.3.2.1 Data processing   Many factors—differences in target hybridization among the arrays, differences in the Cy3 and Cy5 incorporation or degradation rates, and fluorescent intensity variations induced by differences in gain settings when producing images—can complicate the comparison of results from different microarray experiments. To correct for these variations, we applied a global normalization method that multiplicatively normalized the background-corrected spot intensities for each channel of each array to set the 75th percentile equal to 1000. In many experiments, this is nearly equivalent to setting the median of expressed genes equal to 1000. Assuming that most genes are not differentially expressed and that the numbers of overexpressed and underexpressed genes are about the same, this method is also equivalent to the common method used to normalize the fluorescence signal intensities by setting the median ratio between the channels equal to 1. In our case, plots of the log ratio against the mean log intensity of every spot, also known as an M vs. A plot (see Chapter 6), have suggested that this procedure corrected adequately for differences between channels.

Because most of the blank spots on the array had normalized signals below 150, we regarded this as the threshold level and replaced spots with intensity levels below 150 with spots with a value of 150. In addition, we found that the threshold value of 150 corresponded roughly to a spot showing an S/N ratio equal to 1.0 on these arrays, and any spot with a background-corrected intensity below this threshold could not be reliably distinguished from the background noise. We then log-transformed (base 2) the background-corrected intensities for data analysis.

3.3.2.2 Data analysis   Dual-channel fluorescence cDNA microarray data contain a wide range of signal intensities. Using the fold difference between the two channels to identify differentially expressed genes is not entirely reliable, however, because it does not account for the variability in the signal intensity (see Chapter 6). In particular, it is more difficult to determine differential expression of low-intensity genes because these measurements of signal intensity show greater variation, which can be ascribed primarily to background noise. In this study we applied a statistical method that identified differen-

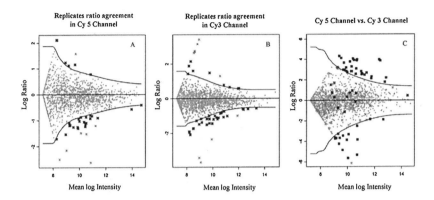

Fig. 3.3 Loess fit to the standard deviation of the replicate ratios within channel on an array produced with the regular protocol. A and B: average log intensity vs. log ratio within each channel based on replicated genes. The asterisks represent poor replicate genes and the bands correspond to ±3 SD. C: Cy5 channel vs. Cy3 channel. The squares outside the bands are differentially expressed genes. The bands represent pooled standard deviations. This figure is reproduced from Wang *et al.* (2003b) with permission from *Biotechniques.*

tially expressed genes on the basis of the studentized log ratio. We describe this approach briefly here.

Recent publications on microarray data analysis have shown that the SD of the log ratio of intensities varies as a function of the mean log-transformed signal intensity. In our method, we first used replicate spots to estimate both the mean log intensity and the SD of the log intensity of the genes within a single channel. (On our array, every spot has been printed in duplicate.) We did this by fitting a smooth (lowess) curve for each channel that described the SD as a function of the mean log intensity. We then took the two channels to be independent, so the variance of the log ratio, $\text{var}(\log(A/B)) = \text{var}(\log(A) - \log(B))$, could be estimated as the sum of the variances of the log intensities, $\text{var}(\log(A)) + \text{var}(\log(B))$. We next pooled the two smooth curves giving the within-channel estimates to obtain a common estimate of the SD of the log ratio between the channels.

Figure 3.3 shows how these smooth curves are applied. In Figure 3.3A and B we examine the agreement between replicate spots within each channel. The log ratio between duplicate spots of the same clone is plotted on the vertical axis, while the mean log intensity of the duplicates is plotted on the horizontal axis. The curves added to the graph represent three times the lowess fit of the SD of the log intensity within that channel; duplicate pairs whose log ratio falls outside these bounds are flagged as poor replicates. In Figure 3.3C, we plot the log ratio between channels vertically and the mean log intensity horizontally. In this graph, the curves superimposed on the graph represent three times the pooled estimate of the SD; points falling outside these bounds represent genes that are differentially expressed.

To determine a statistical significance between the differentially expressed genes, we divide the log ratio of the two channels by the pooled SD to compute a studentized log ratio for each gene as follows:

$$\log R_{\text{studentizied}} = \frac{\log_2(A) - \log_2(B)}{\sigma_{\text{pooled}}}$$

where $\log_2(A)$ and $\log_2(B)$ are the log-transformed background-corrected intensity of each gene in each channel, respectively. This process produces locally studentized values and represents a more robust way to assess differentially expressed genes.

To determine how well an amplification protocol works, we focused on the following issues: (1) the enhancement of the signal intensity, (2) the consistency and reliability of the signal intensity, (3) array reproducibility, and (4) the ability to detect differential gene expression.

3.3.2.3 Enhancement of signal intensity To evaluate the enhancement of signal intensity, we quantified and compared the number of genes with a detectable signal intensity on the amplified and unamplified arrays. Because amplified arrays should produce more spots with an adequate signal intensity, we assessed this quality using the S/N ratio—requiring a spot to have an S/N ratio greater than 2 to be deemed measurable.

	Cy5 (K562 Cells)		Cy3 (RKO Cells)	
	Spots with S/N > 2 (of 2304 spots)	%	Spots with S/N > 2 (of 2304 spots)	%
R-1	704	30.6	1352	58.7
R-2	763	33.1	1770	76.8
A-S1	1483	64.4	2029	88.0
A-S2	1468	63.7	1988	86.3
A-T1	1169	50.7	1877	81.5
A-T2	1238	53.7	1919	83.3

Table 3.1 Results of signal enhancement analysis. S/N: signal/noise (ratio). R-1 and R-2: arrays produced by conventional protocol. A-S1 and A-S2: arrays produced by second strand amplification protocol. A-T1 and A-T2: arrays produced by template switching amplification protocol.

We assessed all six arrays produced by the three different protocols, and the results are summarized in Table 3.1. Far more genes with an adequate signal intensity in both channels resulted from the amplified as opposed to the unamplified protocols. Additionally, more spots with a sufficient signal intensity in both the Cy3 and Cy5 channels were produced by the second-strand cDNA synthesis amplification protocol than by the template-switching protocol.

3.3.2.4 Consistency and reliability of signal intensity To determine whether the amplification protocols preserved the gene signals, we first determined all the spots in each channel that consistently had an S/N ratio > 2.0 on both arrays produced using the regular protocol. We then computed the percentage of those genes that also had an S/N ratio > 2.0 on each set of the arrays produced using the two different amplification protocols. The results are summarized in Table 3.2. Over 93% of the genes detected on the arrays produced using the regular protocol could also be detected on the arrays produced using the two amplification protocols. However, a slightly higher percentage of agreement was seen on the arrays produced using the second-strand cDNA synthesis than on those produced using the template-switching amplification protocol.

Array	Cy5 (K562)	Cy3 (RKO)
Second-Strand Synthesis Amplification		
A-S1	98.9%	99.3%
A-S2	98.6%	99.3%
Template Switching Amplification		
A-T1	93.3%	96.9%
A-T2	95.9%	97.4%

Table 3.2 Results of consistency of signal intensity.

3.3.2.5 Array reproducibility We assessed the reproducibility of arrays by computing the concordance correlation coefficient (Lin, 1989) (r_c) both between the log ratio values and between the studentized log ratio values. Although similar to the Pearson correlation coefficient, the concordance correlation coefficient specifically measures how well points follow the identity line (of perfect agreement) instead of more general linear relations. The reproducibility of the unscaled log ratios and the studentized log ratios within each protocol is shown in Figure 3.4. The results of both analyses showed the high reproducibility of arrays produced using the same protocol.

In terms of different protocols, the r_c between the studentized log ratios for the regular and the second-strand cDNA synthesis amplification protocols

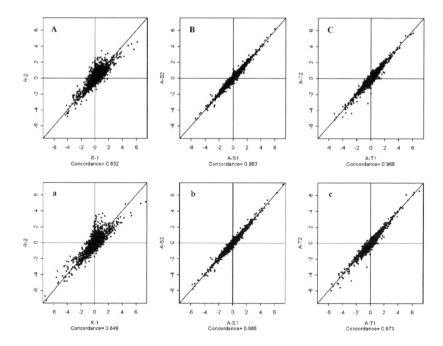

Fig. 3.4 Reproducibility of microarray experiments (A) between arrays produced by the regular protocol; (B) between arrays produced by the second-strand synthesis amplification protocol; and (C) between arrays produced by the template-switching amplification protocol. Panels A–C show the same information, but with studentized log ratios. This figure is reproduced from Wang *et al.* (2003b) with permission from *Biotechniques.*

ranged from 0.690 to 0.816 across the four arrays, with a median value of 0.753. The r_c value between the regular and the template-switching amplification protocols ranged from 0.656 to 0.772 across the four arrays, with a median value of 0.711. The r_c between the second-strand cDNA synthesis and the template-switching amplification protocols ranged from 0.857 to 0.872 across the four arrays, with a median value of 0.867. The concordances between the unscaled log ratios were similar but slightly higher.

3.3.2.6 Ability to detect differentially expressed genes To identify differentially expressed genes, we computed a single studentized log ratio for each gene from duplicated microarrays produced using the same protocol of the three protocols used. We considered genes to be differentially expressed if the combined studentized log ratio exceeded a significance threshold of studentized log ratio > 3.0. Using this cutoff value, we found that 46 genes were differentially expressed between K562 and RKO cell lines in the microarrays

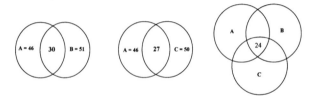

Fig. 3.5 Venn diagrams of differentially expressed genes found on each array, and the genes commonly determined between two different protocols and across three protocols. A: regular protocol; B: second-strand synthesis amplification protocol; C: template-switching amplification protocol. This figure is reproduced from Wang *et al.* (2003b) with permission from *Biotechniques.*

produced using the regular protocol. The microarrays produced using the second-strand cDNA synthesis amplification protocol showed 51 differentially expressed genes, with 30 genes in common with those identified by the regular protocol. Fifty differentially expressed genes were shown by the microarray produced by the template-switching amplification protocol with 27 genes in common with those identified by the regular protocol. Microarrays produced by the three protocols shared 24 genes in common. The results of this analysis are displayed in Venn diagrams (Figure 3.5). For every gene shown to be differentially expressed by at least one of the protocols, the sign of the studentized log ratio value (which determines whether the gene was overexpressed or underexpressed) was the same no matter which protocol was used (Figure 3.6).

Because one of the principal purposes of microarray technology is to identify differentially expressed genes, it is important to determine how well alternative protocols as well as the standard protocol accomplish this goal. Because amplification introduces inherent complications into this assessment, efficiency of amplification may well differ from gene to gene, so we cannot count on the relative intensities of gene expression to remain fixed for a single sample. Fortunately, the efficiency should be the same for a given gene across samples, so one expects the (log) ratios between samples to remain the same. The accuracy of microarray measurements of the log ratio, however, is a function of the mean log intensity (see Chapter 6), and that is why we used studentized log ratios and not simple fold differences to identify differentially expressed genes. Moreover, the expression of some genes that are expressed at a low intensity may be measured more accurately following amplification.

Assuming that the microarrays produced using the regular protocol identified all the differentially expressed genes, we found approximately 65% (30/46) and 59% (27/46) agreement between them and the microarrays produced by the second-strand cDNA synthesis and template-switching amplification protocols, respectively. In terms of the identification of differentially expressed

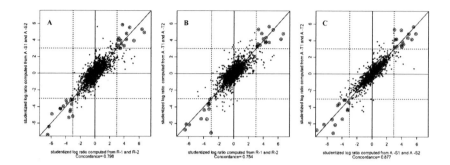

Fig. 3.6 Illustrations of differentially expressed genes identified using two different experiment protocols. (A) Between the regular and second-strand synthesis amplification protocols; (B) between the regular and template-switching amplification protocols; and (C) between the second-strand synthesis and template-switching amplification protocols. The differentially expressed genes in each illustration are indicated in circles. The dotted lines correspond to the *t*-statistic (*t*-score) cutoff value of ±3.0. This figure is reproduced from Wang *et al.* (2003) with permission from *Biotechniques*.

genes, the two amplification protocols performed about the same. However, consistently, more spots with detectable signal intensities in both channels were seen in microarrays produced by the second-strand cDNA synthesis protocol than those produced by the template-switching amplification protocol. The second-strand cDNA synthesis amplification protocol also produced more aRNA after RNA amplification. This is a big advantage for studies for which there is very limited RNA, particularly from tissue samples from patients with rare diseases, such as a rare tumor.

3.3.3   Other amplification strategies

The two protocols described involve the labeling of cDNA after RT. With the availability of cyanine dye-labeled UTP, RNA can also be labeled at the *in vitro* transcription level and used for hybridization. The advantages of this are that it reduces one step in the experimental procedure, and RNA-DNA binding is more stable than DNA-DNA binding. However, the Cy-UTP is not as consistently available as Cy-dCTP is at present.

3.3.4   Selection of method

The method of selection also depends on the type of microarrays to be used. For cDNA microarrays, both strands of the cDNA probes are available for hybridization, and thus labeled targets representing either strand will work. However, for oligo arrays, only one strand of the probes is available for hybridization. Most oligo microarrays use the sense strand of the gene, and thus only the labeled antisense cDNA or RNA can be used for hybridization. Investigators using oligo microarrays should pay special attention to how these microarrays were designed, because use of the wrong stands as targets will generate artifacts or no data. This also highlights the importance of microarray designers providing the sequence information to the end users, who may want to use different hybridization protocols. Problems with this were pointed out in an article in *Nature* (410:860–861, 2001), in which it was reported that Affymetrix had used the wrong strand for building a mouse microarray and that many experiments were wasted as a result.

3.4 INDIRECT LABELING

The discussion up until now has focused on amplification strategies involving the direct labeling of the transcripts and issues involved in the evaluation of new experimental methods. There are also, however, a few amplification methods that use indirect-labeling strategies. When using these methods, similar evaluations should be done to make certain that the data acquired are specific and accurate. In indirect labeling, the fluorescent dyes are not incorporated into the targets used for hybridization; rather, the unlabeled targets are recognized, and thus the signals are detected, by a second hybridization step. Indirect methods were developed to eliminate the problems with the inefficient incorporation of cyanine dye-labeled nucleotides into the DNA or RNA by reverse transcriptase or RNA polymerase as the result of the large size of the cyanine dye molecules. In other words, the processivity of the enzymes is poor when cyanine dye-labeled nucleotides are used, resulting in short products and reduced yields. The indirect method makes the first hybridization step more efficient.

One approach is to use the much smaller aminoallyl nucleotide analogs (aminoallyl-dUTP or aminoallyl-UTP; Molecular Probes, Eugene, Oregon) in the RT reaction or Eberwine amplification reaction. These smaller analogs can be readily incorporated into cDNA, unlike the bulky cyanine dye-labeled nucleotides. The amine group in aminoallyl can then react with cyanine derivatives and more efficiently couple the dye directly to the cDNA or RNA ('t Hoen *et al.*, 2003).

Another indirect-labeling method is used in a technology commercialized by NEN Life Science Products (Boston, Massachusetts) called tyramide signal amplification (TSA), which can amplify the signal 100-fold. In this method, biotin and dinitrophenol (DNP)-labeled nucleotides are first incorporated into cDNA during RT. The cDNA is then hybridized to a microarray and incubated with a horseradish peroxidase (HRP)-conjugated antibody to biotin or DNP. In the following reaction, HRP, in the presence of hydrogen peroxide (H_2O_2), oxidizes the Cy3- or Cy5-tyramide, which then reacts and couples to the probes on the microarray surface, thereby generating fluorescent signals.

3DNA Array Binding

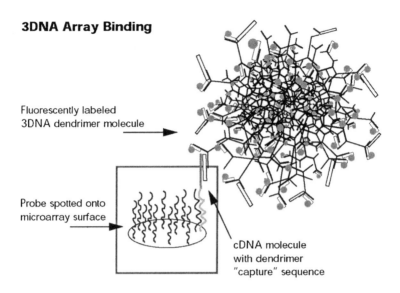

Fluorescently labeled
3DNA dendrimer molecule ⟶

Probe spotted onto
microarray surface ⟶

cDNA molecule
with dendrimer
"capture" sequence

Fig. 3.7 Microarray signal amplification by fluorescent-dye-coupled 3DNA dendrimer. Courtesy of Dr. Ryan Dittamore of Genisphere, Montvale, New Jersey. See insert for a color representation of this figure.

However, to have both Cy5 and Cy3 signals for the two separate samples, two separate hybridization steps are required, and thus these experiments are time-consuming and tedious. However, an advantage to the technique is that a small amount of material is needed for a microarray experiment (Karsten *et al.*, 2002). Researchers also need to keep in mind when using this method that the linearity of the signal amplification may not always be consistent.

Perhaps the cleverest method for indirect hybridization is the 3DNA method commercialized by Genisphere (Hatfield, Pennsylvania). This method utilizes a highly branched nucleic acid structure termed *dendrimer* (Figure 3.7). In an article in the *Journal of Theoretical Biology*, Nilsen *et al.* (1997) laid the foundation for applying dendrimers to DNA detection. As Nilsen explained: "Due to the relatively large size of the nucleic acid molecules, nucleic acid dendrimers can be readily labeled with numerous fluorescent compounds and/or protein moieties with limited steric hindrance and/or quenching." Dendrimers are constructed from unique nucleic acid monomers that contain a stem central region and four single-stranded arms. Subsequent hybridization of the arms results in the expansion of the complex with a growing number of arms. In particular, after the first round of binding, there are 12 arms; after the second round, there are 36 arms; and after the fourth round, there are 324 arms; two more rounds would generate a dendrimer with 2916 single-stranded arms. In addition, for a four-layer dendrimer, one molecule can incorporate more than 300 fluors during production, which provides 300 times the amplification of signals during microarray hybridization (Figure 3.7). A further advantage of this approach is that different dendrimer molecules can have unique sequences at the arms and incorporate different fluorescent dyes, thus enabling multiple samples to cohybridize in one experiment. To use the dendrimer in a microarray experiment, dendrimer-binding sequences are linked to the oligo-dT primers, much like what is done with the T7 promoter sequence in the Eberwine procedure. The cDNA products from RT thus possess the dendrimer binding sequence, which can be detected by the fluorescent-labeled dendrimer complex during the second step of hybridization. Although the first generation of the product did not work well and most signals after hybridization came from nonspecific binding, the newest generation of the product has been improved tremendously. Satisfactory signals are yielded by both cDNA and oligo microarrays using 5–20 μg of total RNA without other amplification steps. This product, therefore, promises to be tremendously useful for microarray detection.

3.5 AUTOMATION OF MICROARRAY HYBRIDIZATION

Currently, researchers perform most hybridization manually, which requires good technical skills and patience. A further concern is that because the volume of the hybridization solution is very small in most cases in which glass-based microarrays are used, the flow of targets between the coverslip and slide is limited and most of the targets present in the buffer do not encounter the probes. Another common problem is that the edge of the slides sometimes dries up during hybridization, even in a humidified oven, which generates a high-background signal on the edge of the microarrays, making the hybridization uneven and the data less reliable.

Several commercial vendors have therefore attempted to design an automated hybridization station that also has a built-in microfluidic agitation feature to facilitate the flow of the hybridization solutions. However, some models of these stations do not perform well because the slide holders leak or cause the slides to break if the slides are held too tightly. Another problem is that some stations require the use of coverslips, which means that when the slide and coverslip are tightened to prevent leakage, a negative pressure forms between them. When the microarrays are disassembled, the DNA probes can be torn off the slides.

One of the more successful automated hybridization stations has been developed by Vantana Medical Systems, Inc. (Tucson, Arizona). It uses a thin layer of inert oil on top of the hybridization buffers to alleviate the negative pressure (Figure 3.8). The microfluidic feature also improves the movement of the targets (Liu *et al.*, 2003). In addition, the hybridization quality is gener-

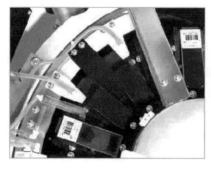

Air Vortex Mixing　　　**Liquid Coverslip™ (LCS) Dispense**

Fig. 3.8 The Air Vortex Mixers stream air to generate a vortex in the aqueous phase via the liquid coverslip for even distribution of reagents and probe over the sample, thus increasing the kinetics of the process and minimizing nonspecific binding. The Liquid Coverslip nozzle dispenses an oil-based coverslip (unlike traditional coverslips) onto the slides, allowing sealing of the sample at all temperatures, thus preventing evaporation during the process. Courtesy of Anis H. Khimani, Ph.D. of Ventana Medical Systems, Inc.

ally uniform and commonly free of an uneven background resulting from dry sides of the hybridization areas on the slides.

3.6 HYBRIDIZATION REFERENCE

Performing cohybridization when using glass-based microarrays allows two samples to be compared directly head-to-head, often enabling one to see differentially expressed genes even before a formal statistical analysis is done. However, because the Cy3 and Cy5 dNTP could be incorporated into some genes at different rates (Liang *et al.*, 2003), dye swap experiments are often performed in order to avoid dye bias (see Chapter 6). Cohybridization also serves as a quality control measure, because as very often happens, when the hybridization is not even across the entire slide, one knows immediately that the results are unreliable. It is for this reason that a reference sample has been widely used for one channel and all the testing samples are in the other channel, which are normalized with the reference channel first before the testing samples are compared with each other. With the automation of hybridization and uniform results, however, reference samples as a hybridization control will probably cease to be needed. Of note, membrane-based microarrays and Affymetrix genechips do not have a cohybridization feature, and one hybridization experiment analyzes only one sample. The advantage of stable single-channel data is that one can get information about relative gene expression levels within one cell population, which is more biologically informative than the ratios calculated on the basis of sometimes totally irrelevant reference sample pools. The fact that two independent samples can be processed on one slide is also cost-effective for microarray analysis because in one hybridization experiment series, two sets of samples can be processed simultaneously on two different channels using one set of arrays. In the future, with additional dyes becoming available, it will be possible to process more than two experiments simultaneously, which will increase the throughput of these already high-throughput technologies.

Although a number of different reference samples have been used in microarray studies, they generally belong to one of three classes. The first class, being the most standard one, consists of RNA-based reference samples. The reference sample can be total RNA from a mixture of cell lines [such as the universal reference RNA mixture from 10 cell lines originally developed in Stanford and commercialized by Strategene (La Jolla, California)] or from cell lines or tissues selected by each investigator. The second class of reference samples is a genomic DNA reference. John Quakenbush's group compared these two classes of reference samples and found that although DNA can serve as a reasonable reference, a much better correlation with direct measurements is achieved by an RNA reference (Kim *et al.*, 2002). The third class of reference is the complementary oligonucleotide sequence for the PCR primers used for the generation of PCR products on microarrays. An advantage of this common reference, which we term OligoRef, is its stability and ability to bind to all the probes on the microarray, thereby eliminating the need to compute ratios against zero or very low values from the reference RNA pools. Several studies, including ours (Hu *et al.*, 2002a; Dudley *et al.*, 2002), have shown that this also serves as a reasonable and stable reference. Sterrenburg *et al.* (2002) used a reference sample made up of a mix of PCR products that are spotted onto the array. However, because all the PCR products contain the same primer sequences, this reference is essentially the same as the oligo reference.

3.7 VALIDATION OF MICROARRAY EXPERIMENTS

Microarray technologies have been very powerful in identifying disease markers, which often can serve as the springboard for long-term mechanism-based investigations. In general, for experimental sciences, results should be confirmed by more than one method, and microarray is no exception. Indeed, due to the complexities involved in microarray production, hybridization, imaging, and data analysis, the validation of microarray data is very important. Another consideration from the standpoint of disease markers is that microarrays can only provide data at the transcript, not at the protein level. For those researchers working in the areas of disease diagnosis and drug development, protein information is more relevant. Therefore, microarray experiments should be done in concert with other assays when a potential target has been selected on the basis of microarray data. Most of the time, this is a very logical and economical approach for research projects. In particular, because microarray experiments are still quite expensive, even when the microarrays are produced by institutional in-house facilities, a relatively small microarray experiment could be performed for screening purposes; then, interesting results could be validated using more conventional and cheaper assays performed on a large number of samples.

There are several validation methods that are commonly used. Real-time PCR assays represent the most popular and efficient method for confirming RNA results. They are used routinely by both basic science and clinical researchers. Northern blotting is often used for cell line-based studies in which the amount of material is not limited. Other RNA-based assays include *in situ* hybridization on a frozen tissue section (Mills *et al.*, 2001), which are able not only to confirm gene expression, but also provide spatial information regarding which cell types in the tissues express the particular transcript. This is an important biological issue, as discussed in Chapter 1. Tissue can be highly heterogeneous, with clinical samples frequently including multiple cell types. However, because microarray data represent population-averaged results, it provides no information regarding the specific sources of cells for expression. The *in situ* hybridization method is especially of value in this regard when there is no antibody available for the protein product of the gene under study.

For clinical studies such as those focusing on marker identification, information at the protein level is often more relevant than information at the RNA or DNA level. Microarrays are very useful as a screening method to identify a marker from which a protein-based pathological assay can be developed. In most hospitals and research institutions, paraffin-embedded tissues, but not frozen tissues, are abundantly available. Therefore, evaluation and confirmation at the protein level can be done with immunohistochemistry analyses as long as a suitable antibody is either available or can be generated. High-throughput tissue microarrays was developed for this purpose (Kononen *et al.*, 1998). In it, 0.6-2-mm cores from hundreds of different tissues can be used to make a recipient block in an organized manner. There are many articles describing such tissue microarrays (see some of the references at the end of this chapter), and thus we will not elaborate on them here. Using tissue microarrays, one hybridization experiment can yield information on hundreds of tissues. For example, our microarray profiling of 25 glioma samples showed that insulin-like growth factor binding protein-2 (IGFBP2) and vascular endothelial growth factor (VEGF) are overly expressed in advanced stages of gliomas. To confirm these findings, we generated a tissue microarray consisting of 256 different glioma tissues of various grades. Staining with anti-IGFBP2 and anti-VEGF antibodies confirmed this result at the protein level (Figure 3.9; Wang *et al.*, 2002a). Further functional studies of IGFBP2

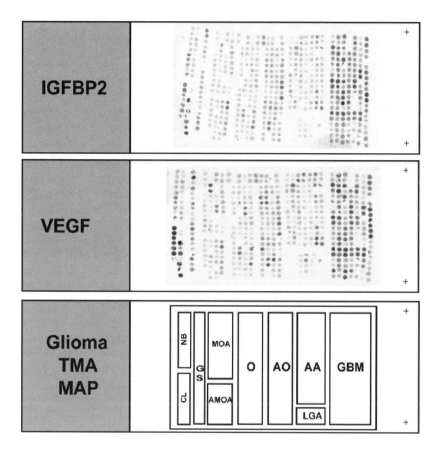

Fig. 3.9 Immunostaining of IGFBP2 and VEGF on a glioma tissue microarray. The bottom panel illustrates the core map of the array, where the acronyms indicate various tumor subtypes. See insert for a color representation of this figure.

revealed that this molecule enhances glioma cell invasion by activating a series of invasion-enhancing genes (Wang et al., 2003a). This is a typical example of the role of expression microarrays in the hypothesis-generating paradigm (Zhang *et al.*, 2002).

3.8 SUMMARY

Many new hybridization protocols are being developed that aim to provide stronger signals from a minute amount of materials. It is always important, however, to evaluate the specificity and sensitivity of any new protocol. Statistical analysis should also be carried out in order to evaluate the performance of a new procedure. Finally, other assays, such as real-time PCR and tissue microarray analysis, should be performed for validating the microarray results and obtaining additional information, such as where a particular gene or protein is expressed in the tissue.

REFERENCES

1. Adler K, Broadbent H, Garlick R, Khimani A, Mikulskis A, Rapiejko P, Killian J. (2000) MICROMAX^{TM}: a highly sensitive system for differential gene expression on microarray. In *Microarray Biochip Technology*, M. Schena, ed. Eaton Publishing, Natick, MA, pp. 221–230.

2. Baggerly KA, Coombes KR, Hess KR, Stivers DN, Abruzzo LV, Zhang W. (2001) Identifying differentially expressed genes in cDNA microarray experiments. *J Comp Biol* 8:639–659.

3. Chuaqui RF, Bonner RF, Best CJ, Gillespie JW, Flaig MJ, Hewitt SM, Phillips JL, Krizman DB, Tangrea MA, Ahram M, Linehan WM, Knezevic V, Emmert-Buck MR. (2002) Post-analysis follow-up and validation of microarray experiments. *Nat Genet* 32 Suppl: 509–14.

4. Clark EA, Golub TR, Lander ES, Hynes RO. (2000) Genomic analysis of metastasis reveals an essential role for RhoC. *Nature* 406:532–535.

5. DeRisi J, Penland L, Brown PO, Bittner ML, Meltzer PS, Ray M, Chen Y, Su YA, Trent JM. (1996) Use of a cDNA microarray to analyse gene expression patterns in human cancer. *Nat Genet* 14:457–460.

6. Dudley AM, Aach J, Steffen MA, Church GM. (2002) Measuring absolute expression with microarrays with a calibrated reference sample and an extended signal intensity range. *Proc Natl Acad Sci USA* 99:7554–7559.

7. Dudoit S, Yang YH, Callow MJ, Speed TP. (2002) Statistical methods for identifying differentially expressed genes in replicated cDNA microarray experiments. *Statistica Sinica* 12:111–139.

8. Duggan DJ, Bittner M, Chen Y, Meltzer P, Trent JM. (1999). Expression profiling using cDNA microarrays. *Nat Genet* 21:10–14.

9. Eisen MB, Brown PO. (1999) DNA arrays for analysis of gene expression. *Methods Enzymol* 303:179–205.

10. Fuller GN, Rhee CH, Hess KR, Caskey L, Wang R, Bruner JM, Yung WKA, Zhang W. (1999) Reactivation of insulin-like growth factor binding protein 2 expression in glioblastoma multiforme: a revelation by parallel gene expression profiling. *Cancer Res* 59:4228–4232.

11. Golub TR, Slonim DK, Tamayo P, Huard C, Gaasenbeek M, Mesirov JP, Coller H, Loh ML, Downing JR, Caligiuri MA, Bloomfield CD, Lander ES. (1999) Molecular classification of cancer: class discovery and class prediction by gene expression monitoring. *Science* 286:531-537.

12. Hu L, Cogdell DE, Jia YJ, Hamilton SR, Zhang W. (2002a) Monitoring of cDNA microarray with common primer target and hybridization specificity with selected targets. *Biotechniques* 32:528, 530–2, 534.

13. Hu L, Wang J, Baggerly K, Wang H, Fuller GN, Hamilton SR, Coombes KR, Zhang W. (2002b) Obtaining reliable information from minute amounts of RNA using cDNA microarrays. *BMC Genomics* 3:16.

14. Karsten SL, Van Deerlin VM, Sabatti C, Gill LH, Geschwind DH. (2002) An evaluation of tyramide signal amplification and archived fixed and frozen tissue in microarray gene expression analysis. *Nucleic Acids Res* 30:e4.

15. Kim H, Zhao B, Snesrud EC, Haas BJ, Town CD, Quackenbush J. (2002) Use of RNA and genomic DNA references for inferred comparisons in DNA microarray analyses. *Biotechniques* 33:924–30.

16. Kononen J, Bubendorf L, Kallioniemi A, Barlund M, Schraml P, Leighton S, Torhorst J, Mihatsch MJ, Sauter G, Kallioniemi OP. (1998) Tissue microarrays for high-throughput molecular profiling of tumor specimens. *Nat Med* 4:844–847.

17. Liang M, Briggs AG, Rute E, Greene AS, Cowley AW Jr. (2003) Quantitative assessment of the importance of dye switching and biological replication in cDNA microarray studies. *Physiol Genomics* 14(3):199–207.

18. Lichter P. (2000) New tools in molecular pathology commentary. *J Mol Diag* 2:171–173.

19. Lin L. (1989) A concordance correlation coefficient to evaluate reproducibility. *Biometrics* 45:255–268.

20. Liu RH, Lenigk R, Druyor-Sanchez RL, Yang J, Grodzinski P. (2003) Hybridization enhancement using cavitation microstreaming. *Anal Chem* 75:1911–1917.

21. Lou L, Salunga RC, Guo H, Bittner A, Joy KC, Galindo JE, Xiao H, Rogers KE, Wan JS, Jackson MR, Erlander MG. (1999) Gene expression profiles of laser-captured adjacent neuronal subtypes. *Nat Med* 5:117–122.

22. Manduchi E, Scearce LM, Brestelli JE, Grant GR, Kaestner KH, Stoeckert CJ Jr. (2002) Comparison of different labeling methods for two-channel high-density microarray experiments. *Physiol Genomics* 10:169–79.

23. Mills JC, Roth KA, Cagan RL, Gordon JI. (2001) DNA microarrays and beyond: completing the journey from tissue to cell. *Nat Cell Biol* 3:e175–8.

24. Newton MA, Kendziorski CM, Richmond CS, Blattner FA, Tsui KW. (2001) On differential variability of expression ratio: improving statistical inference about gene expression changes from microarray data, *J Comp Biol* 8:37–52.

25. Nilsen TW, Grayzel J, Prensky W. (1997) Dendritic nucleic acid structures. *J Theor Biol* 187:273–284.

26. Perou CM, Jeffrey SS, Rijin M, Rees CA, Eisen MB, Ross DT, Pergamenschikov A, Williams CF, Zhu SX, Lee JCF, Lashkari D, Shalon D, Brown PO, Botstein D. (1999) Distinctive gene expression patterns in human mammary epithelial cells and breast cancers. *Proc Natl Acad Sci USA* 96:9212–9217.

27. Phillips J, Eberwine JH. (1996) Antisense RNA amplification: a linear amplification method for analyzing the mRNA population from single living cells. *Methods* 10: 283–288.

28. Rocke DM, Durbin B. (2001) A model for measurement error for gene expression arrays. *J Comp Biol* 8:557–569.

29. Ross DT, Scherf U, Eisen MB, Perou CM, Rees C, Spellman P, Iyer V, Jeffrey SS, Rijn MV, Waltham M, Pergamenschikov A, Lee JCF, Lashkari D, Shalon D, Myers TG, Weinstein JN, Botstein D, Brown PO. (2000) Systematic variation in gene expression patterns in human cancer cell lines. *Nature Genet* 24:227–235.

30. Sallinen SL, Salllinen PK, Haapasalo HK, Helin HJ, Helen PT, Schrami P, Kallioniemi OP, Kononen J. (2000) Identification of differentially expressed genes in human gliomas by DNA microarray and tissue chip techniques. *Cancer Res* 60: 6617–6622.

31. Stears R, Robert C, Getts C, Gullans SR. (2000) A novel, sensitive detection system for high-density microarrays using dendrimer technology. *Physiol Genomics* 3:93–99.

32. Sterrenburg E, Turk R, Boer JM, van Ommen GB, den Dunnen JT. (2002) A common reference for cDNA microarray hybridizations. *Nucleic Acids Res* 30:e116.

33. Taylor E, Cogdell D, Coombes K, Hu L, Ramdas L, Tabor A, Hamilton S, Zhang W. (2001). Sequence verification as quality control step for production of cDNA microarrays. *Biotechniques* 31:62–65.

34. 't Hoen PA, de Kort F, van Ommen GJ, den Dunnen JT. (2003) Fluorescent labelling of cRNA for microarray applications. *Nucleic Acids Res* 31:e20.

35. Van Gelder RN, von Zastrow ME, Yool A, Dement WC, Barchas JD, Eberwine JH. (1990) Amplified RNA synthesized from limited quantities of heterogeneous cDNA. *Proc Natl Acad Sci USA* 87:1663–1667.

36. Wang E, Miller LD, Ohnmacht GA, Liu ET, Marincola FM. (2000) High-fidelity mRNA amplification for gene profiling. *Nat Biotechnol* 18:457–459.

37. Wang H, Wang H, Zhang W, Fuller GN. (2002a) Tissue microarrays: applications in neuropathology research, diagnosis, and education. *Brain Pathol* 12:95–107.

38. Wang J, Coombes KR, Baggerly K, Hu L, Hamilton SR, Zhang W. (2002b) Statistical considerations in the assessment of cDNA microarray data obtained using amplification. In *Computational and Statistical Approaches to Genomics.* W. Zhang and I. Shmulevich, eds. Kluwer Academic Publishers, Norwell, MA, pp. 23–39.

39. Wang H, Wang H, Shen W, Huang H, Hu L, Ramdas L, Zhou YH, Liao WS, Fuller GN, Zhang W. (2003a) Insulin-like growth factor binding protein 2 enhances glioblastoma invasion by activating invasion-enhancing genes. *Cancer Res* 63:4315–21.

40. Wang J, Hu L, Hamilton SR, Coombes KR, and Zhang W. (2003b) Evaluating the performance of two RNA amplification strategies for cDNA microarray experiments. *Biotechniques* 34:394–400.

41. Yang YH, Dudoit S, Lu P, Speed T. (2001) Normalization for cDNA microarray data. In *Microarrays: Optical Technologies and Informatics* (M.L. Bittner, Y. Chen, A. N. Dorsel, and E. R. Dougherty, eds.), *Proceedings of SPIE,* Vol. 4266, SPIE, Bellingham, WA, pp. 141–152.

42. Zhang W, Wang H, Song SW, Fuller GN. (2002) Insulin-like growth factor binding protein 2: gene expression microarrays and the hypothesis-generation paradigm. *Brain Pathol* 12:87–94.

4

Scanners and Data Acquisition

The purpose of this chapter is to provide a general understanding of the microarray scanning process. The emphasis is on general principles and their effect on the quality of the acquired microarray image, which, in turn, is the input to the image analysis software. This software then performs quantification or, in other words, estimation of the abundance of the target RNA species in the sample. From a practical point of view, it is not necessary for a biologist performing microarray analysis to know the details of scanner designs. It is, however, useful to have an intuitive understanding of the sources of noise and distortions that are unavoidable in electro-opto-mechanical systems.

Each microarray experiment consists of a long chain of delicate steps from tissue acquisition to microarray data analysis, and typically, errors and distortions can get magnified in each step. A basic understanding of the scanner technology may minimize problems during these phases and help in the design of experiments such that the full potential of the imaging phase is utilized. The devices that read microarray slides or chips and output an image or images can be divided into two classes—scanners and imagers. Scanners typically use laser excitation and a confocal microscope to scan (with high speed) the slide a very small area at a time. An imager works like a digital camera, using white light as an excitation to take snapshots over large portions of the slide. These are then stitched together to obtain the microarray image.

4.1   BASIC PRINCIPLES OF SCANNERS

cDNA microarrays consist of thousands of spots of small amounts of pure nucleic acid species printed in a high-density array on a glass microscope slide using a robotic printing system. Fluorotagged representations of mRNA of several (usually, two) cell types, each emitting light of a different wavelength, are jointly hybridized to the array. The relative abundances of the mRNAs in samples for each spot can then be read with an imaging device capable of detecting the intensities of the components of light emitted by the different fluorescent molecules, each within its specific wavelength range.

A typical system uses two samples labeled with the fluorescent dyes Cy3 (green fluorescent) and Cy5 (red fluorescent). The spots are printed on the slide in a rectangular array that is often composed of subarrays or subgrids. The scanner forms two high-resolution images of the slide, one for Cy5 and one for Cy3, where the brightness of each pixel reflects the amount of Cy5 or Cy3 in the area corresponding to the pixel. Often, these images are combined, placing each image in the appropriate color channel of an RGB image. Intense red color then indicates a high level of Cy5, whereas intense green signifies a high level of Cy3.

Each spot consists of a number of pixels (typically, 50–400), collectively representing the total amounts of Cy5 and Cy3, with the corresponding gene expression levels to be estimated by the image analysis software. The relation between the amounts of labeled mRNA and the pixel intensities produced by the scanner is very complicated, and many factors contribute to this relation via different mechanisms, resulting in a highly nontrivial correspondence.

The quality and accuracy of signal measurement can be degraded by various sources of noise and distortions, which can be divided into three general categories. The first group of sources are those in reading the intensity of a single pixel by laser-excited confocal microscope; the second are those resulting from the necessity of the scanner to move either the slide or the microscope or both with high speed to cover the entire spotted area in a reasonable time. As the number of pixels is on the order of millions, very high mechanical precision is required. Third, because of the many physical and chemical factors involved, even in ideal conditions, the spots are highly variable, and depending on the image analysis software, different types of distortions are either introduced or remain uncorrected during the estimation process. The sources of noise can also be divided into instrument noise, which is produced by the imaging system itself, and microarray noise, which results either from substrate, coating, and sample or by nonspecific hybridization to the probes on the microarray surface.

To appreciate the difficulties in reading the amounts of the fluorophores, consider the structure of a laser-excited confocal microscope typically found in microarray scanners and depicted in Figure 4.1. To be able to detect the small quantities of photons emitted by the fluorescent molecules, photomultiplier tubes (PMTs) are used to convert the light energy into electric current that

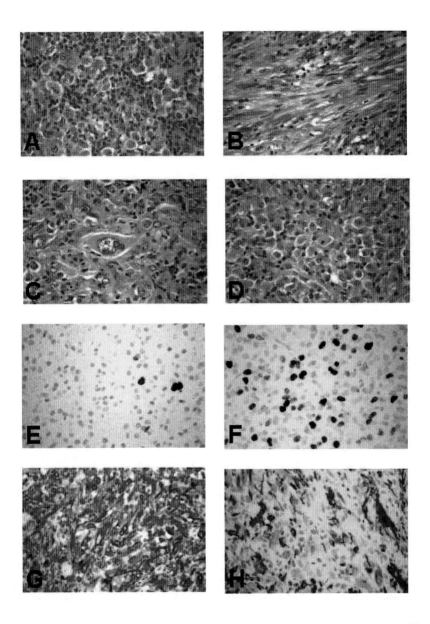

Fig. 1.1 Phenotypic, proliferative, and molecular heterogeneity in brain tumors. Phenotypic heterogeneity (A–D): Four microscopic fields from the same glioblastoma show small cell (A), spindle cell (B), giant cell (C), and epithelioid (D) differentiation. (H&E, 200×.) Proliferative heterogeneity (E–F): Two different areas of the same oligodendroglioma show strikingly different proliferation activities. (MIB-1 immunocytochemistry; 200×.) Molecular heterogeneity (G–H): Two areas of one glioblastoma illustrate variable GFAP gene expression. (GFAP immunocytochemistry; 200×). (This figure was generously provided by Dr. Gregory N. Fuller, The University of Texas M. D. Anderson Cancer Center.)

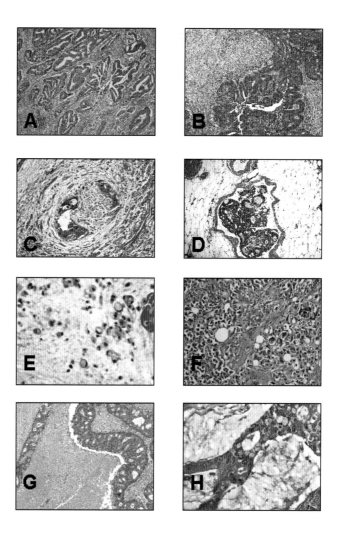

Fig. 1.2 Phenotypic heterogeneity in colon tumors. Moderately differentiated adenocarcinoma of colon (A–D): Elongated malignant glands arranged in a corkscrew pattern invade into the stroma (A). A different tumor displays a cribriform pattern and a prominent periglandular lymphocytic infiltrate (B). (H&E, 200×.) Same tumor focally invades the perineurium (C) and is present in the vascular space (D). (H&E, 400×.) Poorly differentiated adenocarcinoma of colon (E–F): Discohesive signet-ring cells floating in pools of mucin (E). (H&E, 400×.) Highly anaplastic tumor with pleomorphic cells, invading into the pericolonic adipose tissue (F). (H&E, 200×.) Adenocarcinoma of colon with ribbonlike pattern (G–H): Enlarged, dilated malignant glands with central necrosis (G). (H&E, 200×.) Mucinous tumor with pleomorphic cells and abundant extracellular mucin (H). (H&E, 400×.) (This figure was generously provided by Dr. Lucian Chirieac, The University of Texas M. D. Anderson Cancer Center.)

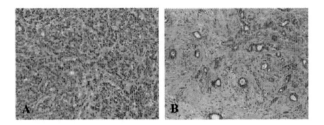

Fig. 1.3 The histopathological appearance is variable among different invasive breast cancers. Some cancers consist of an almost pure population of breast cancer cells (A). Most cancers contain a significant component of desmoplastic stroma (B) that is active in angiogenesis, fibrosis, myocontractility, and immunity. (This figure was generously provided by Dr. Fraser Symmans, The University of Texas M. D. Anderson Cancer Center.)

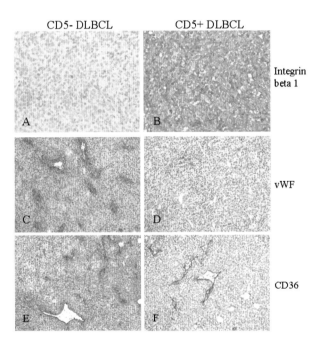

Fig. 1.4 Immunohistochemical staining of lymphoma tissues for integrin beta 1, vWF, and CD36. Integrin beta 1 was weakly stained for CD5-negative DLBCL (A) and strongly stained for CD5-positive DLBCL (B). vWF was stained strongly in the vascular cells of both CD5-negative and CD5-positive DLBCL (C and D). CD36 was not stained in CD5-negative DLBCL (E) and stained highly in the vascular cells of CD5-positive DLBCL (F).

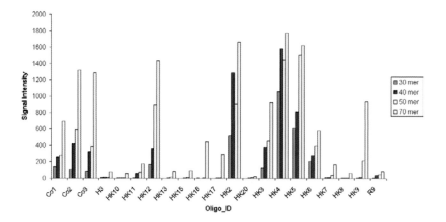

Fig. 2.1 The effect of oligo length on detected signal intensities. For each gene evaluated, the signal intensities from the four probes of different length are plotted on the bar graph.

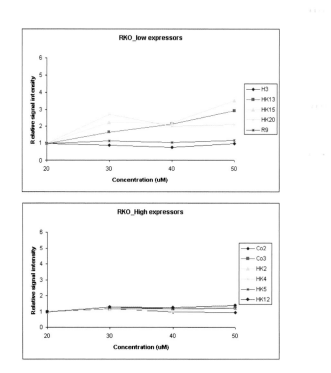

Fig. 2.2 The effect of probe amount on detected signal intensities. Relative intensities for 20, 30, 40, and 50 μM concentration of the probe in comparison to 20 μM.

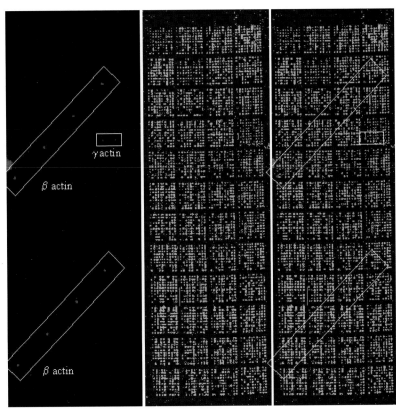

A   **B**   **C**

Fig. 3.2 Specific hybridization of β-actin target on the array. A β-actin target was generated by RT-PCR from K562 leukemia cell RNA. Cy5-dCTP was incorporated during PCR. The downstream primer used to generate all cDNA on the array, except for GAPDH and β-actin, was end-labeled with Cy3. Cy5-labeled β-actin and Cy3-labeled primer were cohybridized to the microarray slides. The array contains 2303 known human genes printed in replicate, in addition to positive and negative controls. After washing and scanning of the slide, the hybridization result showed that (A) the β-actin target was specifically hybridized to the corresponding spots on the array and the (B) Cy3-labeled primer hybridized to every spot generated by PCR using the same primer, indicated by green spots on the array. (C) The composite image of both Cy3 and Cy5 channels. Hybridization to β-actin and γ-actin was indicated. This figure is reproduced from Hu *et al.* (2002a) with permission from *Biotechniques*.

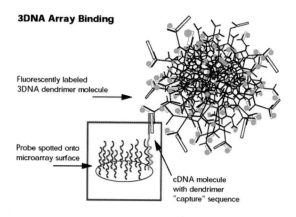

Fig. 3.7 Microarray signal amplification by fluorescent-dye-coupled 3DNA dendrimer. Courtesy of Dr. Ryan Dittamore of Genisphere, Montvale, New Jersey.

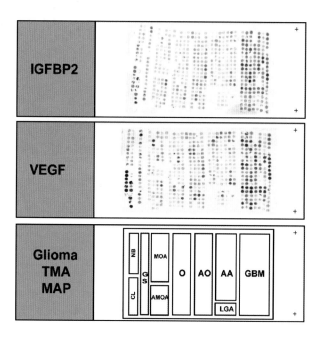

Fig. 3.9 Immunostaining of IGFBP2 and VEGF on a glioma tissue microarray. The bottom panel illustrates the core map of the array, where the acronyms indicate various tumor subtypes.

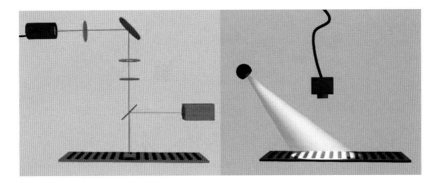

Fig. 4.1 Simplified pictures of a microarray scanner (left) and imager (right).

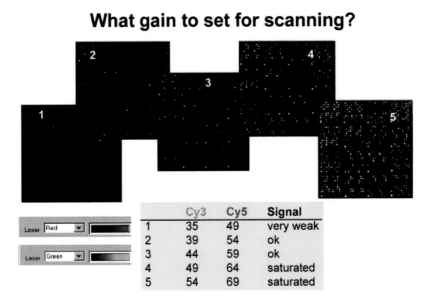

Fig. 4.2 An illustration of the effect of different gain settings on the resulting microarray image. A low gain results in very few detectable spots, whereas a high gain causes many spots to saturate (shown as white), essentially making them indistinguishable and negatively affecting downstream data analysis. (This figure was provided by Dr. Latha Ramdas, The University of Texas M. D. Anderson Cancer Center.)

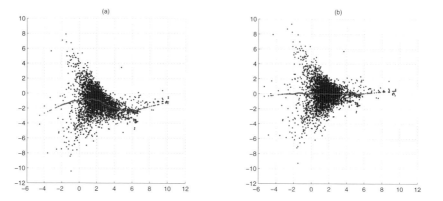

Fig. 6.2 An M vs. A plot of the same data set as in Figure 6.1, shown in panel (a). The lowess curve, with 20% of the data used at each point, is overlaid on the plot and is used to normalize the data. The results of the normalization are shown in panel (b). As a confirmation, the lowess curve estimated from the normalized data is nearly a straight line.

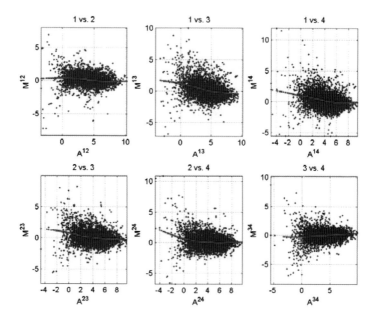

Fig. 6.4 The six scatter plots show all combinations of $M_i^{(p,q)}$ vs. $A_i^{(p,q)}$ using the Cy5 channel of four different cDNA microarrays. The fitted lowess curves clearly indicate a need for intensity-dependent normalization.

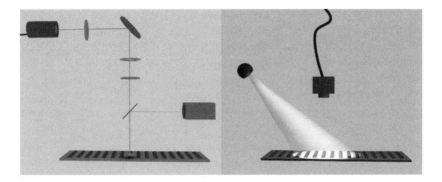

Fig. 4.1 Simplified pictures of a microarray scanner (left) and imager (right). See insert for a color representation of this figure.

can be measured. In a PMT, incoming photons hit a photosensitive electrode, which emits electrons that travel inside the tube and hit additional electrodes, generating yet more electrons. This process is repeated in the tube several times, resulting in very large amplification of the initial number of electrons. The amplification can be controlled by the voltage applied to the tube. The resulting current is proportional to the incident light at the photocathode. PMTs are very sensitive, fast, and fairly linear over their operating range.

The scanner repeats the detection of the signal for each pixel and dye. Some scanners read the entire row of pixels for one dye and then switch the laser and repeat the process for another dye. Yet other scanners gate the lasers and read both (all) dyes at the same location before moving to the next pixel. As the pixel size is only 5–10 μm, high mechanical precision is needed.

The optical path from the slide back to the detector needs to maximize energy collection. This calls for increasing the numerical aperture of the optical system, to let in more light. However, this decreases the depth of field. It is important that the scanner depth of field be correct so that it is not collecting emissions from below or above the spot surface. To handle the variations in slide surface, some scanners employ adaptive adjustments for the distance between the microscope and the slide surface, making millions of corrections during one scan.

4.2 BASIC PRINCIPLES OF IMAGERS

A microarray imager generates a digital image of the microarray hybridization result in a cameralike fashion. Typically, imagers use white light that is generated by a powerful xenon arc lamp. An excitation filter passes the wavelengths that are needed to excite the particular probes, and the fluorescence emission is passed through the emission filter and captured by a charge-coupled device (CCD) camera.

The imager captures a large portion (1 cm^2 at a time) and the entire image is formed by tiling these subimages together. Careful overlapping of several subimages can be used to reduce instrument noise. This architecture is simpler and more flexible than the scanner architecture. However, a scanner can deliver more excitation photons to the probes than an imager, thus generating more photons per pixel per unit of time. Also, imagers are less sensitive than PMTs and the imager's wideband excitation source is less efficient. Consequently, it takes longer for the camera to capture the fluorescent signals.

Another potential problem is cross-talk between different imager channels, as it is more difficult to obtain pure excitation and emission signals with white light than with laser-based systems. Also, the area sensors used in imagers typically generate much more instrument noise than PMTs and need to be cooled to obtain sufficiently low dark count to produce the desired dynamic range.

4.3 CALIBRATION AND PMT GAIN

To produce meaningful data the scanner needs to be calibrated, which involves setting the laser power and the PMT voltages such that the response to a known fluorescence standard is a specified signal. High-end scanners provide automatic calibration of lasers. The PMT operates as a very high-gain amplifier to the signal emitted by the fluorescent molecules. Its response is quite linear over a large range, and ideally, the PMT voltage is adjusted so that the signal occupies the full dynamic range of the sensor. A gain that is set too low increases quantization noise and makes weak spots difficult to analyze. On the other hand, an excessively high gain causes a range of high signal values to saturate, thus making them indistinguishable, and raises the background values. Figure 4.2 illustrates this phenomenon.

For the typical case of two colors, one can calibrate either both channels or the channel ratio. In light of the discussion above, a preliminary individual calibration of the channels followed by minimal adjustments to achieve calibration of the channel ratio should be optimal. Additional difficulties arise from the fact that the PMT has slightly different responses to the wavelengths corresponding to the colors.

Optimally, the PMT gain should be set such that the full dynamic range is utilized. Increasing the gain generally does not increase the signal-to-noise ratio (SNR) because the noise becomes amplified more than the signal. At the same time, lowering the gain below this optimal value also does not increase the signal-to-noise ratio because the photon detection is less efficient at low gain levels. Figure 4.3 illustrates the effect of increasing the gain on the SNR. The five panels show histograms of the SNRs above 2 for various gain values

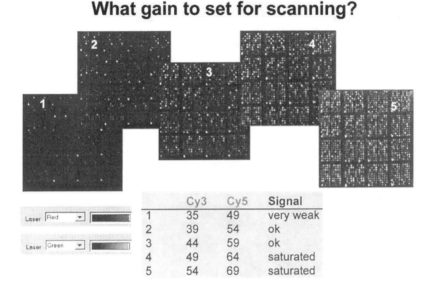

Fig. 4.2 An illustration of the effect of different gain settings on the resulting microarray image. A low gain results in very few detectable spots, whereas a high gain causes many spots to saturate (shown as white), essentially making them indistinguishable and negatively affecting downstream data analysis. (This figure was provided by Dr. Latha Ramdas, The University of Texas M. D. Anderson Cancer Center.) See insert for a color representation of this figure.

between 34 and 54. It can be seen that as the gain is increased, there are fewer spots with high SNR values.

An alternative approach is *line averaging,* or repeatedly scanning the same slide and averaging over scans, which increases the signal-to-noise ratio in the square root of the number of scans. Dudley *et al.* (2002) used a linear regression algorithm to combine the linear ranges of multiple scans taken at different scanner sensitivity settings onto an extended linear scale. The drawbacks of doing this are possible photobleaching and increased complexity and handling time.

4.4   CHARACTERISTICS OF DIFFERENT NOISE SOURCES

A microarray scanner is a complex instrument containing electrical, optical, and mechanical parts. Consequently, there are many sources of distortions that affect the final microarray image. In the following, we consider briefly the

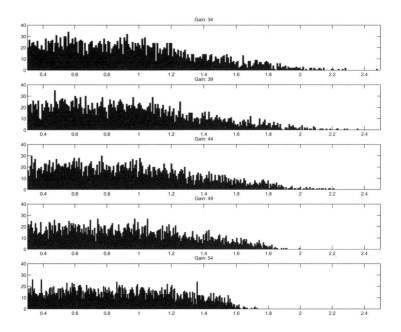

Fig. 4.3 Histograms of $\log_{10}$-transformed signal-to-noise ratios (SNRs) for various PMT gain settings. The gain values used (top to bottom) are 34, 39, 44, 49, 54. Only the SNRs above 2 (0.3010 on the $\log_{10}$ scale) are shown. The scanner used was the GeneTAC LSIV scanner (Genomic Solutions, Ann Arbor, Michigan).

main types of noise appearing in microarray images and discuss the fundamental limitations that affect calibration. We can distinguish between background noise and signal noise.

4.4.1 Background noise

4.4.1.1 Dark current The current that appears without excitation is called dark current and results from thermal emissions. Figure 4.4 shows a typical image formed from dark current only. The level of dark current, measured in electrons per second, is often called the dark count. To reduce dark count, some systems employ cooling that keeps the sensors at temperatures below $-50°C$.

4.4.1.2 Shot noise The detectors in microarray scanners work on the level of detecting individual photons, and thus there exist intrinsic physical limitations stemming from the particle nature of light. The number of molecules hybridized on an area in the spot is governed by Poisson statistics, where the

Fig. 4.4   A typical dark-field image that is formed from the dark current only.

variance is proportional to the square root of the power of the signal. Likewise, the power of noise in the emitted signal, shot noise, is proportional to the square root of the power of the signal. This relationship implies that if the signal is increased by integrating over more detected photons, the overall signal-to-noise ratio decreases.

4.4.1.3 Laser noise   The amount of light emitted is proportional to the amount of fluorophore molecules and the amount of laser light applied to the area. Thus, all variations in the laser light affect the signal directly. Typically, the drift in laser intensity is more serious than the noise in the laser, and some scanners have internal control mechanisms to eliminate the drift adaptively. Also, part of the excitation light makes its way to the sensor and causes additional noise.

4.4.1.4 PMT noise   The PMT detects photons emitted by the fluorophore molecules and creates greatly amplified signal by a series of electron emissions. The variations in the resulting signal that are not caused by variations in initial photon emission are called PMT noise. This noise is multiplicative with respect to the original signal. Figure 4.5 contains an example of multiplicative noise.

4.4.1.5 Quantization noise   The digitizing step always produces quantization noise. As 16-bit quantization is generally used, yielding 65,536 different signal levels, in a well-designed scanner quantization noise is negligible. However, if

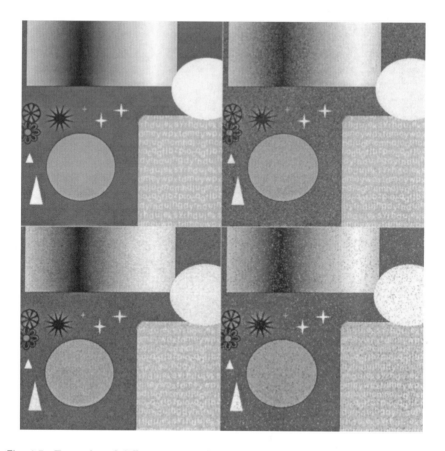

Fig. 4.5 Examples of different types of noise. Original image (upper left); additive Gaussian white noise (upper right); multiplicative or *speckle noise* (bottom left); and spike noise (bottom right). Note that the multiplicative noise, unlike additive noise, is image dependent, which is why in the darker regions of the image, corresponding to smaller pixel values, the noise power is also smaller. In particular, the totally black regions contain no multiplicative noise. Correspondingly, the noise is stronger in bright image areas. To generate spike noise, both white and black pixels were randomly placed in the image. We chose to use an artificial image to illustrate the effects of the noise, since it is not possible to obtain a real microarray image that is completely noise-free.

the signal occupies only a small portion of the dynamic range of the quantizer, severe quantization noise might occur. Figure 4.6 illustrates the effect of quantization on an image.

4.4.1.6 Surface variability Microarray slides have surface variability that is on the order of the thickness of the spot, and unless the scanner can compen-

Fig. 4.6 The effect of quantization noise. The image on the left is the original 8-bit image (256 grayscale levels). The image on the right is the image quantized to 3 bits (8 grayscale levels). No dithering was used.

sate for this variability, it needs to have larger depth of field. This implies that the sensor will pick up more photons emitted below the surface, leading to slowly varying background noise.

4.4.1.7 Spike noise Dust and other defects on the surface produce spike noise, that is, small spots of high intensity. Also, microscopic scratches or small pieces of fabric dust produce bright snakelike distortions. Figure 4.5 shows examples of different types of noise, including spike noise.

4.4.1.8 Nonspecific binding Residues of the fluorescent materials may bind to the surface of the slide, causing an additional component in the background noise (also see Chapter 2).

4.4.1.9 Local contamination In addition to these distortions in the signal, there are several kinds of artifacts produced during printing, hybridization, and washing stages that affect parts of the slide and may render some areas useless. Figure 4.7 shows an example of local contamination.

4.4.2 Signal noise

All forms of background noise are also present within the spots, but because the signal is much stronger, only spike noise and local contamination are clearly distinguishable. There are many forms of distortion in the spots, but these are less like noise in nature and more appropriate to consider in the image analysis stage (see Chapter 5).

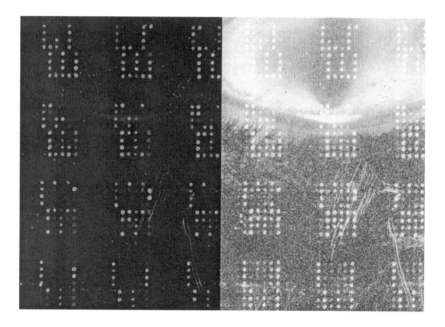

Fig. 4.7 An example of local contamination. Although the contamination is present in the original image (left), it is not visible, due to poor contrast. After applying histogram equalization, the contamination becomes apparent (right).

REFERENCES

1. Dudley AM, Aach J, Steffen MA, Church GM. (2002) Measuring absolute expression with microarrays with a calibrated reference sample and an extended signal intensity range. *Proc Natl Acad Sci USA.* 99(11):7554–9.

2. Kamberova G, Shah S. (2002) *DNA Array Image Analysis: Nuts & Bolts.* DNA Press, Skippack, PA.

3. Schena M. (2003) *Microarray Analysis.* John Wiley & Sons, New York.

4. Schena M. (1999) *DNA Microarrays: A Practical Approach*, 2nd ed. Oxford University Press, Oxford.

5. Schena M., ed. (2000) *Microarray Biochip Technology.* Eaton Publishing, Natick, MA.

6. Ramdas L, Wang J, Hu L, Cogdell D, Taylor E, Zhang W. (2001) Comparative evaluation of laser-based microarray scanners. *Biotechniques* 31:546-553.

7. Ramdas L, Coombes KR, Baggerly K, Abruzzo L, Highsmith WE, Krogmann T, Hamilton SR, Zhang W. (2001) Sources of nonlinearity in cDNA microarray expression measurements. *Genome Biol* 2(11): research0047.

8. Van den Doel LR, van Vliet LJ, Young IT. (2001) Design considerations for a conventional microscope based microarray-reader to improve sensitivity, *Proceedings of the Second Euroconference on Quantitative Molecular Cytogenetics*, Salamanca, Spain, pp. 108–113.

5

Image Analysis

As in chapter 4, the purpose here is to provide a general understanding of microarray image analysis. We again emphasize general principles and their subsequent effect on data analysis. As for the case of scanner technology, it is not necessary for a biologist performing microarray analysis to know all the details of the image analysis algorithms. However, it is useful to have an intuitive understanding of how the algorithms estimate the expression levels for being able to assess the suitability of a particular software or an option therein for the task at hand as well as for appreciating how these choices affect the quality of the analysis. It is clear that the better the image, the easier the image analysis task. For perfect spots, even the simplest algorithm works well, and on the other hand, no image analysis can reproduce the information that has been lost in the previous steps by, for example, the wrong setting in the scanning process.

The effects of various algorithms and methods on the quality of the image and data analysis can be studied by employing simulated data sets. The basic idea, commonly used in the signal and image processing community, is to create an artificial microarray image that closely resembles the features and statistics of real microarray images. Various contaminations, distortions, and noise can be added to this image and the image analysis algorithms can be assessed and compared on the basis of how well the true information is estimated or recovered. Several groups have carried out these kinds of studies (Balagurunathan *et al.*, 2002; Wierling *et al.*, 2002).

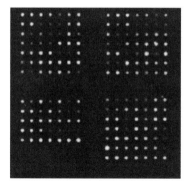

Fig. 5.1   Example of subgrids of irregular size.

5.1   GRID ALIGNMENT

The structure of a microarray image is determined by the printing process, and thus it is known in advance. Typically, the spots are printed in subgrids that contain a certain number of rows and columns. The subgrids also form a regular array—the metaarray. A microarray image showing the metaarray of subgrids is displayed in Figure 2.6.

The purpose of the gridding process is to locate the subgrids in a microarray image and place grid lines so that the spots lie within the crossings of the grid lines. This would give the coordinates of the exact locations of the spots for subsequent analysis. Ideally, all subgrids are of the same size, the spacing between subgrids is regular, and the distance between adjacent rows and columns in subgrids is constant. In practice, however, this is not the case and the metaarray can be far from regular, with the distance between subgrids varying quite a bit (Figure 5.1).

The subgrids may be tilted by several degrees, which makes algorithms assuming purely horizontal and vertical lines prone to errors. Also, the distance between spots in subgrids can vary substantially (Figure 5.2), and after placing the grids, the locations of spots must also be estimated. There are several levels of sophistication in gridding.

Manual gridding. The simplest approach is to manually place the grids with a software tool that allows the user to point to the corners and the endpoints of the lines. A drawback is that it is very time consuming and also prone to human errors as well as loss of consistency due to different operators.

Semiautomatic gridding. The second level is semiautomatic gridding, where the operator frames the grid area using a mouse tool while the software estimates and shows the gridlines and gives the operator the opportunity to adjust the lines. Contrast enhancement and other image-processing opera-

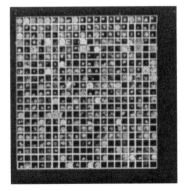 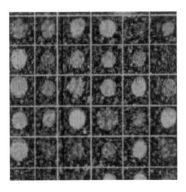

Fig. 5.2 A gridded image (left) and an example where the exact locations of spots with respect to the grid may vary considerably (right).

tions that improve the visual quality of the image are performed routinely to aid in placing the grid.

Automatic gridding. The best solution is a fully automatic and accurate algorithm that can place the grid quickly and accurately. Let us briefly consider the methods that have been developed for this purpose. Parts of these algorithms are also employed in semiautomatic methods.

If full printing information is available, then the software knows fairly accurate initial estimates of the subgrid locations and only fine adjustments are needed. This can be obtained with a type of matched filter, which roughly means that the software computes pixel-by-pixel correlations with a smaller model image having the same line spacings. To be effective, the line spacings in the model image should be very close to those of the image analyzed and its orientation (tilt) should be the same. Estimation of the spacings and orientation can be done in several ways. Fourier transform methods can be utilized to speed up matching operations. Some algorithms obtain the line spacing by taking horizontal and vertical projections of the grid. This works very well for untilted grids or when several projections over a range of orientations are performed. This effectively also reveals the orientation. It is important to note that many imaging operations, such as evaluating different projections, which are simple tasks for a human, are nontrivial to do automatically. Performing the gridding separately for the red and green channels can reveal misregistration of the two channels and allow for its correction.

After gridding, the spots lie approximately within the cross-points of the grid. Owing to many imperfections in the printing process, there is always some variation in the exact spot locations, and sometimes this variability can be quite large. The spot-finding process, which can be considered either the final task of gridding or the first task of the spot segmentation, consists of specifying the exact coordinates of the spot centers and placing rectangular

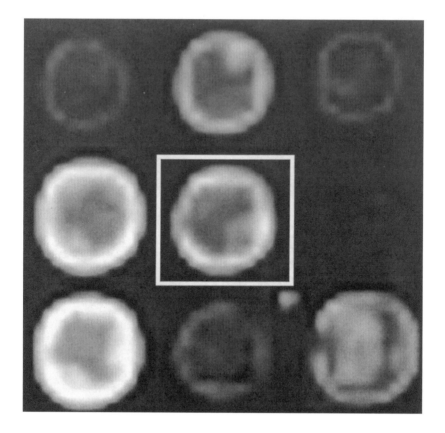

Fig. 5.3 A rectangular patch placed around a spot.

areas (patches) of usually equal sizes around the spots (Figure 5.3). The locations of spots can be found with similar matched filtering as the subgrids, but now a simple bump is fitted in a small circular area around the grid crossing, and the best match is chosen as the spot location.

5.2 SPOT SEGMENTATION

If all spots were of the same size and regular in form, spot segmentation would be an easy task, and the simplest segmentation algorithms are in fact based on this assumption. In reality, the spots vary in several ways. In many instances, because the printing tip touches the glass surface, the spot has a lower intensity or even a hole in the center such that it has more or less a doughnut shape (Figure 5.4). The substance left by the printing tip does not spread evenly to all directions but has an irregular shape. It may also spread

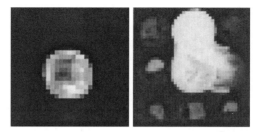

Fig. 5.4 Close-up pictures of spots showing the doughnut-type shape (left) and bleeding (right).

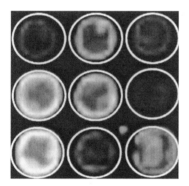 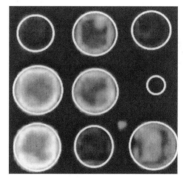

Fig. 5.5 Illustration of fixed circle segmentation (left) and adaptive circle segmentation (right).

extensively and "bleed" into a neighboring spot.

Many algorithms of varying levels of complexity and based on different image-processing principles have been developed. In the following we examine more closely the best known algorithms in order of increasing sophistication.

Fixed circle segmentation. Fixed circle segmentation places a circle of constant size at each spot location (Figure 5.5). The inside of the circle is assumed to contain the spot, and the outside, background. Because the actual spots may not correspond to circles of a particular size, there will be significant errors in the intensity values, due to the fact that background pixels are included in the calculation of intensity of the spot, and vice versa.

Adaptive circle segmentation. In adaptive circle segmentation the radii of the circles are estimated for each spot separately. This provides a better fit but cannot take into account other deformations in spots. GenePix (Axon Instruments Inc.) uses such an approach.

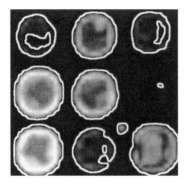 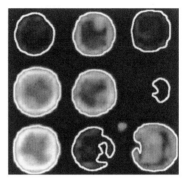

Fig. 5.6   Illustration of intensity-based segmentation (left) and region-growing segmentation (right).

Adaptive shape segmentation. Adaptive segmentation methods aim at estimating the true form of the spot. They fall into two main categories: histogram-based methods and region-growing methods.

In *histogram methods*, the algorithm determines a threshold value and the pixels that have intensity higher that the threshold are classified as belonging to the spot and the rest to the background (Figure 5.6). This method is employed by QuantArray (GSI Luminomics). It is difficult to determine a good threshold value reliably and sophisticated statistical methods have been developed for this task. The segmentation method developed by Chen *et al.* (1997) uses a robust statistical test, the Mann-Whitney test, to determine the threshold. This test is nonparametric, meaning that it works reliably independently of the actual distribution of intensity values. The main drawback in histogram-based methods is that they cannot take shape information into account. Thus, spike noise or contamination from other spots may be classified into the spot, leading to errors in the estimated intensity values.

The *region-growing methods* have their roots in mathematical morphology and typically start with a seed pixel—a pixel that can reliably be assumed to belong to the spot—and add pixels one by one as long as a suitable criterion is satisfied. For nonhomogeneous spots, the selection of the initial seed is not trivial and can result in spurious segmentation. The accuracy can be improved by combining spatial and intensity information using iterative schemes, based on spatial and intensity information, that move pixels back and forth between spot and background regions until a satisfactory balance has been obtained.

Another approach that has been used to segment spots is based on *clustering analysis* (Bozinov and Rahnenführer, 2002; Nagarajan and Peterson, 2002). The idea behind such an approach is to apply some clustering technique, such as *k*-means, to the pixels inside a patch. Some pixels then get assigned to the spot class while the others get assigned to the background class.

5.3   EXTRACTING INFORMATION

After the image has been segmented to spots and background, the next step is to estimate the intensities of the spots and the background. We have seen earlier that the background intensity may vary considerably in different parts of the microarray slide, due to scanner focus differences or other factors. This means that to obtain accurate estimates of the background intensity, local estimates should be used. The distribution of the background intensity can be modeled as a mixture of a Gaussian distribution and impulsive noise.

The simplest estimator is the average value of the background pixels in the patch. However, the problem with the average is that it is sensitive to spike noise, especially because the dynamic range of a microarray image is large. A few spikes can considerably shift the background estimate (Figure 5.7). This can be avoided by using a *trimmed mean*, where a percentage of the largest and smallest values are removed before taking the mean. Another robust estimate is the median value of the background pixels, but for a Gaussian distribution it is theoretically inferior to the trimmed mean.

Using all background pixels of the patch to calculate the background intensity is ideal in the sense of signal-to-noise ratio, which improves when averaging over increasing number of points. The drawback is that when the interspot distance is small, the neighboring spots may bleed into the area over which averaging is done and distort the estimate. This can be avoided by selecting a number of pixels in the corners of the patches for calculation of the background intensity (Figure 5.8). After the background levels have been estimated for all patches, it is reasonable to analyze their variability. Large changes among background levels in patches close to each other may be indicative of a problem and can be used as a quality control measure.

The total amount of hybridization for a spotted DNA sequence is proportional to the total fluorescence in a spot, which means that the simplest measure is the sum of the signal intensities in the segmented spot area. The background level is assumed to be the same within the entire patch and the signal intensity is defined as *signal intensity = spot intensity − background intensity*. To eliminate the effect of variability in spot sizes, typically the average intensity of the spot pixels is used. For the ratio between red and green channels, this is equivalent to the total intensity. The mean is sensitive to outliers, and often, median or trimmed means are used instead. A number of sophisticated statistical models for the channel ratios have been developed. These allow the construction of test statistics for the significance of differential expressions as well as spot quality measures.

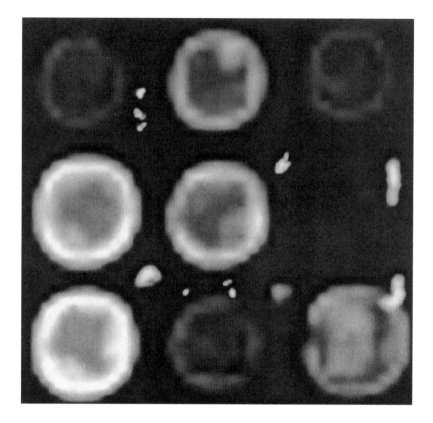

Fig. 5.7   Illustration of spike noise, which can distort background estimates.

5.3.1   Spot quality

There are several features that affect the quality of a spot on a microarray. We will briefly introduce the ones encountered most frequently in microarray experiments. Signal intensity is traditionally considered to be one of the most important features affecting spot quality. The reason is that if a signal is weak, it is very difficult to distinguish the actual signal from the background. Accordingly, spots with higher signal intensity have a higher signal-to-noise ratio and thus are more reliable than spots with low signal intensity. Most weak spots are caused by the fact that most genes are physiologically expressed at very low levels, at or close to the sensitivity limit, with gene expressions in a typical microarray experiment following a lognormal distribution. In addition, experimental factors that may cause low signal intensities are: low amount of DNA in the spot, molecular and physical composition of the DNA spot (purity of the DNA, attachment to glass, availability to hybridization), uneven

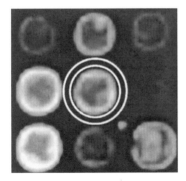 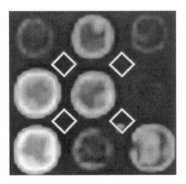

Fig. 5.8 Background estimation from a circular area around the spot (left) and from diagonally placed rectangular patches (right).

labeling of samples with fluorophores, uneven or incomplete hybridization, signal bleaching, and low sensitivity of the scanner.

Another important factor is spot size. As microarray spots are produced by depositing an equal amount of liquid onto the microscope slide, all spots are expected to be of roughly equal size. Deviations can be caused by, for example, precipitates, impurities and debris of the printing solution, a printing pin that is not making adequate contact with the array surface, damaged or dirty pins, or an insufficient amount of liquid printed onto the slide. Spots that are too large can be caused by high humidity during printing or uneven coating of the slide. Both of these can cause the spot to spread over a larger area than intended, which can lead to problems with detection and quantification of surrounding spots. Whether a spot is significantly larger or smaller than expected, it is an indication that there has been some problem during manufacturing, which should be reflected when determining the quality of the spot.

The geometric form of a spot is a factor that also needs to be addressed in assessing its quality. Spots are expected to be roughly circular in shape, but manufacturing of the microarrays may cause some variation of the spot morphology. These include scratches on the microarray surface and uneven coating of the microarray. Other imperfections on the microarray's surface can also cause the spot to spread unevenly during printing. Furthermore, if the signal is weak, the spot may have an irregular shape because spot signal is not distinguished from the background. A spot may also spread so much that it bleeds to the patches of its neighbors. In such cases, it is generally impossible to separate the signals of the spots from each other reliably. Therefore, spots affected by bleeding are often excluded from further analysis. The pixel intensities in a spot should, ideally, exhibit little variation. Deviations from this are visible as brighter areas inside the spot and can be caused by uneven distribution of the printed DNA in the spot or nonspecific binding.

5.4   APPENDIX: IMAGE PROCESSING

The result of microarray hybridization is read by either a scanner or an imager as a rectangular array of fluorescence values. Although the final result of the experiment is a set of estimates of gene expression levels, the reading process generates a high-resolution image of the microarray. Because of this, much information about the experiment can be obtained simply by visual examination of the microarray image. In the quantification stage, one uses the visual image for tasks such as grid placing, while for automatic analysis, many image processing techniques are utilized. It is important to note that the main goal of a microarray scanner or imager is to produce accurate estimates of gene expression levels and to generate an image for human viewing. A consequence of this is that one must be careful in judging quality on the basis of visual inspection. Also, many features in microarray images are not visible to the human eye without some enhancement or modification.

The purpose of this appendix is to recall some basic concepts of image processing and the human vision system from the point of view of microarray images. The fundamental steps in image processing are

- Acquisition

- Storage

- Enhancement and restoration

- Segmentation

- Feature extraction

- Classification and interpretation

- Display

If an image is being processed for human viewing, then feature extraction, classification, and interpretation are typically omitted. Image-processing techniques are best described as signal processing operations and modeled as mathematical algorithms.

5.4.1   Image models

The mathematical model of a monochrome image is a function f of two variables,

$$f(x, y), \quad -\infty < x, y < \infty,$$

where x and y denote spatial coordinates and $f(x, y)$ represents the optical intensity or brightness at location (x, y). Consequently, f takes on only non-negative values. A color image is a collection of monochrome images, each representing the intensity of one particular color (e.g., red, green, and blue).

A digital monochrome image is an image that has been discretized both in spatial coordinates and intensity. It is conveniently represented as a matrix

$$I(m, n), \ m = 1, \ldots, M, \ n = 1, \ldots, N,$$

where the elements of the matrix are called *picture elements* or *pixels,* taking values in a finite range $\{1, \ldots, K\}$. In typical image processing applications, M and N are in the range 200–4000 and $K = 256$. For a monochrome image this means that there are 256 possible gray levels, and for a three-color image that there are about 16 million different colors. These are often referred to as *8-bit gray-level images* or *24-bit color images*, respectively. It is important to note that although the human eye has an enormous dynamic range, on the order of 10^{10}, it is mainly a question of brightness adaptation, and once adapted, the eye can locally (in a small area of an image) distinguish only about 20 gray levels. However, as the eye scans over an image, it adapts continuously and about 200 gray levels are needed to represent a whole image. The situation is similar with color images, and the number of color shades that the eye can distinguish locally is very small compared to the number of possible shades in a 24-bit image. These properties make visual images very different from microarray images, where the dynamic range is usually 16 bits for each channel (Cy3 and Cy5), and for displays they need to be suitably transformed.

An important point in digitizing any physical image is that the spatial resolution of the digitizer must be fine enough compared to spatial variations in the continuous image. Unless this is the case, severe distortions, so called aliasing, can occur in the digitized image. In well-designed optical systems, the optical components and the sensor (having finite area) smooth the signal sufficiently to avoid aliasing. However, the phenomenon can reappear, for example, in resizing, unless adequate smoothing is done prior to the resizing operation.

5.4.2 Image storage

In image processing, a number of standards have emerged for storing digital images. These standards also lend themselves to storing microarray images, although the latter are commonly saved in the tagged image format (TIFF) with 16 bits per pixel. One must be very careful when transferring images between devices, especially of different manufacturers, because a microarray image can be represented as a "visual" image in many ways. Also, various devices can introduce different sources of nonlinearity into the image data (Ramdas *et al.*, 2001).

In addition, compressing microarray images using lossy image compression methods, such as JPG, is not advisable unless the compressed image is for human viewing only. The reason is that lossy compression standards typically discard all information that is not visually noticeable but may be important. Furthermore, such compression standards (e.g., JPG) introduce so-called *blocking artifacts*, which are particularly visible near nonvertical or nonhorizontal edges (sharp transitions), as shown in Figure 5.9.

Fig. 5.9 An example of blocking artifacts due to JPG compression. These artifacts are especially prevalent near nonvertical or nonhorizontal edges (i.e., sharp transitions from black to white).

5.4.3   Enhancement and restoration

Image enhancement and restoration concern reducing the degradations and deformations that are introduced during the formation of the image. In restoration one tries to recover the original ideal image, whereas in enhancement the aim is to improve the quality of the digital image, especially to emphasize those properties of the image that are essential in further processing either by a human or by an automated system. In the following, we briefly consider the most typical enhancement and restoration techniques that are useful in microarray data processing.

Contrast transformation or histogram transformation refers to a pixel-by-pixel mapping of the original gray levels to new ones. There are two main reasons for doing histogram transformations. First, the information in images is often concentrated in a few small ranges of gray levels, and the human eye is not able to resolve these small differences. Stretching the histogram or equalizing it (i.e., making a transformation that spreads the gray level as evenly as possible) improves clarity, and many details may become visible. The second reason is that to operate optimally, many image-processing algorithms assume certain statistical properties of the input signal (i.e., image), such as its distribution being uniform or Gaussian. A histogram transformation can be used to change the input image such that it more closely matches the assumptions of the processing algorithms. One typical example is an image that is corrupted by multiplicative noise; that is, each image pixel is modeled as

$$I(m, n) = I_t(m, n)N(m, n),$$

where $I_t(m, n)$ is the true pixel value and the noise $N(m, n)$ is positive. Taking a logarithmic transform leads to an additive noise model:

$$\log(I(m, n)) = \log(I_t(m, n)) + \log(N(m, n)).$$

Image filtering refers to operations that aim at reducing noise or emphasizing either slow or rapid variations. A typical filtering operation, so-called *low-pass filtering*, which reduces noise by replacing each pixel value by the average value of the nearby pixels in a small window, is expressed mathematically as

$$I_o(m, n) = \frac{1}{(2R + 1)^2} \sum_{i=-R}^{R} \sum_{j=-R}^{R} I(m + i, n + j).$$

It is intuitively clear that for an additive noise model

$$I(m, n) = I_t(m, n) + N(m, n),$$

where the true image $I_t(m, n)$ is slowly varying, the noise will be well reduced and that the averaging has little effect on the true image. A problematic situation that is often present in microarray images is when there are edge-like level changes in the image. Then, the averaging operation will smooth the edge, and for instance, spot segmentation will become more difficult. This problem can be alleviated by using the median operation instead of averaging:

$$I_o(m, n) = \text{MED}\{I(m + i, n + j) : i, j = -R, \ldots, R\}$$

where the median is the centermost value after ordering; for example,

$$\text{MED}\{2, 3, 4, 1, 5\} = 3.$$

The median filter has less smoothing effect on edges, but on the other hand, is somewhat less effective in removing additive Gaussian noise.

To emphasize rapid variations such as edges, high-pass filtering is used. Instead of taking local averages, it typically computes local differences, so the parts of images having rapid variations get high positive or negative values. The same effect can be obtained by subtracting a low-passed image from the original image. This technique is called *unsharp masking*.

When image filtering is needed for visual inspection, most advanced image processing software packages contain all necessary filtering operations. When filtering is needed as a preprocessing step for an automatic system that estimates gene expression levels from microarray images, careful modeling that includes thorough analysis of the noise and distortion sources during the scanning process is required to ensure that the filtering itself does not generate additional errors in the end results.

Image transformations play a central role in signal and image processing. The idea is to represent the image as a sum of simple basis images. The Fourier transform results from representing the image as a sum of two-dimensional complex sinusoids. Let $I(m, n)$, $0 \le m \le M$, $0 \le n \le M$ be an image. Then, its discrete Fourier transform is defined as

$$\hat{I}(s,t) = \sum_{m=0}^{M-1} \sum_{n=0}^{N-1} I(m,n) e^{-j(2\pi/M)sm} e^{-j(2\pi/N)tn},$$

where $0 \le s \le M - 1$, $0 \le t \le N - 1$ and $j = \sqrt{-1}$. When working with transforms it is more convenient to index the pixels starting from 0. Notice that the transform is a complex image of the same size as the original. When displaying the discrete Fourier transform, often only the image consisting of the magnitudes $|\hat{I}(s,t)|$ of the transform values is shown. Because the Fourier transform effectively decomposes the image into periodic components, it very clearly reveals periodicities in the image.

One of the most important features of the Fourier transform is that many filtering operations are significantly faster to perform in the transformed domain. For instance, averaging-type filtering operations performed in the image domain, for each pixel, may require summations and multiplications over large ranges, whereas in the transformed domain, a single multiplication per pixel is sufficient.

In the Fourier transform, the image is decomposed into sinusoidal images of the same size as the original image. The implication is that while the Fourier transform extracts precise information of the frequencies present in the image, it cannot reveal local properties of the frequencies. Another type of transform, the *wavelet transform*, decomposes the image into elementary waves of different scales and location. Thus, it is better able to reveal local properties of the image.

Image segmentation refers to the decomposition of an image into regions of uniform characteristics. It is a key step in image analysis. For instance, a microarray image needs to be segmented into "background" and "spot" before estimating the intensity of the spot. The simplest segmentation method is to divide the image into regions by amplitude thresholding. It is useful when amplitude features sufficiently characterize the images. For more complex images, such as microarray images, more advanced methods are needed. They typically use high-pass filtering to extract edges in the image and segment the image into areas surrounded by edges. This usually results in an oversegmented image (i.e., the regions are too small to represent useful features of the image). A region-merging algorithm analyzes neighboring regions, and if they satisfy the criterion of similarity, merges them into one. Another class of segmentation algorithms is region-growing segmentation methods, whereby each region starts to grow from a seed that is either manually placed or based on prior information. Pixels are added one by one to the regions as long as the criterion characterizing regions stays fulfilled.

Feature extraction is a preprocessing step for classification and analysis. For instance, in character recognition, after the image has been segmented to subregions consisting of characters and their immediate background, a number of numerical values, called *features*, such as the number of strokes, angles between strokes, or even transform coefficients, are computed. The actual classification is then based on these simplified characteristics or features. Typically, in practical algorithms a sparse representation (i.e., a feature space of low dimension) is needed.

Classification and analysis refer to the final steps where an image or region is classified in one of several classes based on the values of the features computed. There is a large number of different classification methods of varying levels of sophistication. The simplest ones are based on computing distances between the feature point of the item to be classified and reference points, while more complex ones combine learning methods and artificial intelligence.

REFERENCES

1. Angulo J, Serra J. (2003) Automatic analysis of DNA microarray images using mathematical morphology. *Bioinformatics* 19(5):553–62.

2. Astola J, Kuosmanen P. (1997) *Fundamentals of Nonlinear Digital Filtering.* CRC Press, Boca Raton, FL.

3. Balagurunathan Y, Dougherty ER, Chen Y, Bittner ML, Trent JM. (2002) Simulation of cDNA microarrays via a parameterized random signal model. *J Biomed Opt* 7(3):507–23.

4. Beucher S, Meyer F. (1993). The morphological approach to segmentation: the watershed transformation. Mathematical morphology in image processing. *Opt Engineering* 34:433–481.

5. Bozinov D, Rahnenführer J. (2002) Unsupervised technique for robust target separation and analysis of DNA microarray spots through adaptive pixel clustering. *Bioinformatics* 18(5):747–56.

6. Brown C, Goodwin P, Sorger P. (2001) Image metrics in the statistical analysis of DNA microarray data. *Proc Natl Acad Sci USA* 98:8944–8949.

7. Burrus CS, Gopinath RA, Guo H. (1997) *Introduction to Wavelets and Wavelet Transforms.* Prentice Hall, Upper Saddle River, NJ.

8. Chen Y, Dougherty ER, Bittner ML. (1997) Ratio-based decisions and the quantitative analysis of cDNA microarray images. *J Biomed Optics* 2(4):364–374.

9. Chen Y, Kamat V, Dougherty ER, Bittner ML, Meltzer PS, Trent JM. (2002) Ratio statistics of gene expression levels and applications to microarray data analysis. *Bioinformatics* 18(9):1207–1215.

10. Dougherty, ER. (1992) *An Introduction to Morphological Image Processing.* SPIE, Bellingham, WA.

11. Glasbey CA, Ghazal P. (2003) Combinatorial image analysis of DNA microarray features. *Bioinformatics* 19(2):194–203.

12. Gonzales RC, Woods RE. (2002) *Digital Image Processing*, 2nd ed. Addison-Wesley, Reading, MA.

13. Hautaniemi S, Edgren H , Vesanen P, Wolf M, Järvinen, AK, Yli-Harja O, Astola J, Kallioniemi O, Monni O (2003): A novel strategy for microarray quality control using Bayesian networks. *Bioinformatics* 19: 2031–2038.

14. Jain, AK. (1989) *Fundamentals of Digital Image Processing.* Prentice-Hall International, Inc., Upper Saddle River, NJ.

15. Jain AN, Tokuyasu TA, Snijders AM, Segraves R, Albertson DG, Pinkel D. (2002) Fully automatic quantification of microarray image data. *Genome Res* 12(2):325–32.

16. Kamberova G, Shah S. (2002) *DNA Array Image Analysis: Nuts & Bolts.* DNA Press, Skippack, PA.

17. Kim JH, Kim HY, Lee YS. (2001) A novel method using edge detection for signal extraction from cDNA microarray image analysis. *Exp Mol Med* 33(2):83–8.

18. Kooperberg C, Fazzio TG, Delrow JJ, Tsukiyama T. (2002) Improved background correction for spotted cDNA microarrays. *J Comput Biol* 9(1):55–66.

19. Nagarajan R, Peterson CA. (2002) Identifying spots in microarray images. *IEEE Trans Nonobiosci* 1(2):78–84.

20. Petrou M, Bosdogianni P. (1999) *Image Processing: The Fundamentals.* John Wiley & Sons, New York.

21. Ramdas L, Coombes KR, Baggerly KA, Abruzzo L, Highsmith WE, Krogmann T, Hamilton SR, Zhang W (2001). Sources of nonlinearity in cDNA microarray expression measurements: *Genome Biol* 2(11):research0047.

22. Ruosaari S, Hollmén J (2002) Image analysis for detecting faulty spots from microarray images. In S. Lange, K. Satoh, C. H. Smith eds., *Proceedings of the 5th International Conference on Discovery Science.* Springer Verlag, New York, pp. 259–266.

23. Schena M, Shalon D, Davis RW, Brown PO. (1995) Quantitative monitoring of gene expression patterns with a complementary DNA microarray. *Science* 270(5235):467–70.

24. Smyth GK, Yang YH, Speed T. (2002) Statistical issues in cDNA microarray data analysis. In *Functional Genomics: Methods and Protocols*, M. J. Brownstein, A. B. Khodursky, and D. B. Conniffe, eds., Methods in Molecular Biology Series. Humana Press, Totowa, NJ.

25. Soille P. (2003) *Morphological Image Analysis: Principles and Applications*, 2nd ed. Springer Verlag, New York.

26. Steinfath M, Wruck W, Seidel H, Lehrach H, Radelof U, O'Brien J. (2001) Automated image analysis for array hybridization experiments. *Bioinformatics* 17(7):634–41.

27. Tran PH, Peiffer DA, Shin Y, Meek LM, Brody JP, Cho KWY. (2002) Microarray optimizations: increasing spot accuracy and automated identification of true microarray signals. *Nucleic Acids Res* 30(12):e54.

28. Tseng GC, Oh MK, Rohlin L, Liao JC, Wong WH. (2001) Issues in cDNA microarray analysis: quality filtering, channel normalization, models of variations and assessment of gene effects. *Nucleic Acids Res* 29(12):2549–57.

29. Wang X, Ghosh S, Guo SW. (2001) Quantitative quality control in microarray image processing and data acquisition. *Nucleic Acids Res* 29(15): e75–5.

30. Wierling CK, Steinfath M, Elge T, Schulze-Kremer S, Aanstad P, Clark M, Lehrach H, Herwig R. (2002) Simulation of DNA array hybridization experiments and evaluation of critical parameters during subsequent image and data analysis. *BMC Bioinformatics* 3(1):29.

31. Yang MC, Ruan QG, Yang JJ, Eckenrode S, Wu S, McIndoe RA, She JX. (2001) A statistical method for flagging weak spots improves normalization and ratio estimates in microarrays. *Physiol Genomics* 7(1):45–53.

32. Yang YH, Buckley MJ, Dudoit S, Speed TP (2002). Comparison of methods for image analysis on cDNA microarray data. *J Computational Graphical Statist* 11:108–136.

6

Quality Control in Data Analysis

Data analysis typically represents the last stage of a microarray experiment. It is at this step that biologically relevant conclusions are typically made. Despite careful and stringent quality control throughout all of the previous stages described thus far, the results of the experiment and ensuing conclusions could become ruined if data analysis is performed improperly. This stage presents a substantial number of difficulties and challenges. The goal of this chapter is to reveal some of these pitfalls and suggest ways to identify and prevent them.

6.1 NORMALIZATION

In a microarray experiment, there are many sources of variation. Some types of variation, such as differences of gene expressions, may be highly informative, as they may be of biological origin. Other types of variation, however, may be undesirable and can confound subsequent analysis, leading to wrong conclusions. In particular, there are certain systematic sources of variation, usually due to specific features of the particular microarray technology, that should be corrected prior to further analysis. The process of removing such systematic variability, called *normalization*, is an important aspect of quality control in microarray data analysis.

There may be a number of reasons for normalizing microarray data. For example, there may be a systematic difference in quantities of starting RNA, resulting in one sample being consistently overrepresented. There may also be differences in labeling or detection efficiencies between the fluorescent dyes (e.g., Cy3, Cy5), again leading to systematic overexpression of one of the samples. Thus, in order to make meaningful biological comparisons, the measured intensities must be properly adjusted to counteract such systematic differences.

Most normalization methods require us to make certain natural assumptions. For instance, we need to assume that the arrayed genes represent a random sample of the genes of the organism being surveyed. Indeed, suppose that our microarray contains only those genes that we know are overexpressed in sample A relative to sample B. Then, we will expect to see a systematic bias that is due to the biology and not the measurement technology, obviating normalization. If we start with equal amounts of RNA obtained from the two samples and we assume that the average mass of each molecule is roughly the same, then the total number of molecules in each sample will also be the same. Consequently, an overrepresentation of one gene in a sample (in terms of the number of molecules) will be offset by an underrepresentation of another gene in the same sample. Therefore, as the total number of molecules is assumed to be the same, the sum of all the measured hybridization intensities should be the same for both samples. This reasoning leads us to *global normalization*.

6.1.1   Global normalization

Let us first consider the case of two samples being cohybridized to the same array, using different fluorescent dyes. This normalization (within-slide global normalization) consists of applying a simple multiplicative constant factor such that the ratio of the sums of the intensities (or equivalently, the arithmetic means) in the two samples is equal to 1. That is, suppose that there are n genes and the measured intensities of the ith gene in the red (Cy5) and green (Cy3) channels are denoted by R_i and G_i, respectively. Let

$$k_{\text{mean}} = (\sum_{i=1}^{n} R_i)/(\sum_{i=1}^{n} G_i)$$

be the ratio of the arithmetic means of the two channels. Global normalization consists of scaling one of the channels, say, the green channel: $G_i' = k_{\text{mean}} \cdot G_i$ and leaving the other channel intact: $R_i' = R_i$. Therefore, the ratio of the means of the normalized channels will be equal to

$$(\sum_{i=1}^{n} R_i')/(\sum_{i=1}^{n} G_i') = (\sum_{i=1}^{n} R_i)/(k_{\text{mean}} \sum_{i=1}^{n} G_i) = 1.$$

On the logarithmic scale,[1] the multiplicative factor amounts to an additive constant. That is, if the ith normalized ratio is denoted by R_i'/G_i', then

---

[1] Logarithm base 2 is commonly used, since twofold overexpression corresponds to +1 and twofold underexpression corresponds to −1.

$$\log_2\left(R_i'/G_i'\right) = \log_2\left(R_i'\right) - \log_2\left(G_i'\right) = \log_2\left(R_i\right) - \log_2\left(G_i\right) - \log_2\left(k_{\text{mean}}\right).$$
Thus, the center of distributions of log ratios becomes shifted to zero.

Instead of using the ratio of the means as a multiplicative constant, other location estimates can be used. A popular approach is based on the *median*[2]:

$$k_{\text{med}} = \text{median}\left(R_1, \ldots, R_n\right)/\text{median}\left(G_1, \ldots, G_n\right).$$

The advantage of the latter approach is that the median is much more robust than the mean in the presence of outliers. In other words, it's value is much less sensitive to extreme values (i.e., very highly expressed genes) than the mean and gives a more realistic estimate of the 'average' array intensity. Despite this, the median may not always be preferred, due to the fact that microarray distributions are often highly skewed, meaning that most genes are expressed at very low levels and a few genes are expressed at very high levels. Indeed, the gene expressions in a microarray typically follow a lognormal distribution[3] (Hoyle *et al.*, 2002). Thus, the medians of the two channels may both be quite low (close to zero). To address this problem, a larger percentile, such as the 75th percentile, may be used.

For purposes of illustration, let us consider an array that exhibits a significant difference between the two channels. The plot of the unnormalized log ratios $M_i = \log_2\left(R_i/G_i\right)$ vs. the mean log intensities $A_i = \frac{1}{2}\log_2\left(R_iG_i\right)$ is shown in Figure 6.1(a). The M vs. A plot is essentially a 45° rotation of the $\left(\log_2 G, \log_2 R\right)$ coordinate system. Such plots are preferred in many studies, as they more naturally reveal intensity-dependent patterns, artifacts, and differential expression. It can easy be seen that there are more data points below zero on the M-axis. This implies that the mean intensity of the red channel is lower than that of the green channel. Indeed, the former is equal to 7.26, while the latter is 21.08; thus $k_{\text{mean}} = 0.34$. The globally normalized data are shown in Figure 6.1(b).

6.1.2   Intensity-dependent normalization

As Figure 6.1 shows, global normalization cannot remove systematic bias completely, even though the average intensities of the two channels are equal. In particular, the scatter plot reveals a certain pattern in the ratios relative to the average log intensity. For example, it appears that as the average log intensity (A) increases from 1 to 5, the log ratios (M) get smaller. Thus, there is a systematic intensity-dependent bias. One way to remove such a bias is to smooth the data with a locally weighted regression method, such as *lowess*. This method is useful for smoothing scatter plots to reveal the underlying

---

[2] The median is the value in a set of ranked observations that divides the set into two equal parts. For an odd number of observations, the median is the middle value. For an even number of observations, it is the average of the two central values.

[3] A random variable X is lognormally distributed if $\ln\left(X\right)$ is normally distributed.

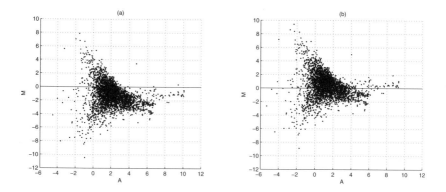

Fig. 6.1 An M vs. A plot of an array that exhibits a significant difference between its two channels. The plot in panel (a) corresponds to the original un-normalized data. The plot in panel (b) corresponds to the globally normalized data ($k_{\mathrm{mean}} = 0.34$).

patterns or structure and for identifying nonlinear relationships, in this case, between M and A. There are robust versions of this procedure that are resistant to outliers. Additionally, only a certain percentage of the data, such as 20%, is typically used for the smoother at each point. Let $l(A_i)$ be the value of the lowess smoother at point A_i. Then, if we let $G'_i = G_i \cdot 2^{l(A_i)}$ and $R'_i = R_i$,

$$
\begin{aligned}
M'_i &= \log_2 (R'_i/G'_i) = \log_2 (R_i) - \log_2 (G_i) - l(A_i) \\
&= \log_2 (R_i/G_i) - l(A_i) \\
&= M_i - l(A_i).
\end{aligned}
$$

Thus, lowess-based *intensity-dependent normalization* consists of simply subtracting the smooth curve from the original log ratio data. Applying intensity-dependent normalization in the M vs. A space is preferable to applying it directly in the $\log_2 (R_i)$ vs. $\log_2 (G_i)$ space, as the former addresses the variability in both channels simultaneously by regressing to their geometric mean.

Consider Figure 6.2(a). The dots represent the original unnormalized data, which is the same as those in Figure 6.1(a). The overlaid curve is the lowess smoothed data, with 20% of the total number of data points used at each point. It can be seen that the shape of this curve confirms our observation that there is a tendency of the log ratios to decrease when the average log intensity ranges from 1 to 5. The dots in Figure 6.2(b) indicate the intensity-dependent normalized data. As a confirmation, another overlaid lowess curve applied to the normalized data is nearly a straight line, indicating a successful normalization.

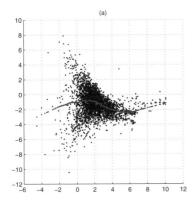 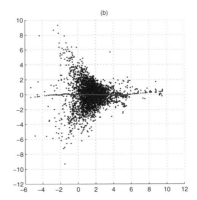

Fig. 6.2 An M vs. A plot of the same data set as in Figure 6.1, shown in panel (a). The lowess curve, with 20% of the data used at each point, is overlaid on the plot and is used to normalize the data. The results of the normalization are shown in panel (b). As a confirmation, the lowess curve estimated from the normalized data is nearly a straight line. See insert for a color representation of this figure.

6.1.3 Pin-dependent normalization

We have now considered two kinds of normalization that can be used for a two-color array. What these two methods have in common is that the normalization factor, be it a constant or a lowess curve, is computed for the entire array. However, in spotted microarray technology, a number of different pins are used to produce the array. The printed spots are usually arranged in *subgrids*, and all spots in a given subgrid are printed with the same pin. For example, suppose that the print head contains 48 pins arranged in a 4×12 layout. Then, the microarray will consist of 48 subgrids, each subgrid containing some number of spots (see Figure 2.6). It is possible that there may be systematic sources of variation, due to the differences between the pins. For example, one pin might be slightly deformed, or in the case of split pins, may have a smaller opening. Accordingly, the spots created with that pin will tend to be different from the spots created by the other pins. Thus, the corresponding subgrid will also have a systematic bias. This leads to the idea of applying a normalization procedure to each pin separately. For example, intensity-dependent normalization, such as the lowess-based approach, can be applied 48 times, each time using only the data from a given pin. This type of *pin-dependent normalization* should be used with care, especially when the number of spots printed by each pin is relatively small, since the observed differences between different pins could be due to chance. In order for us to determine whether there are significant pin-dependent differences in the log ratios, we could do the following, which we illustrate with an example.

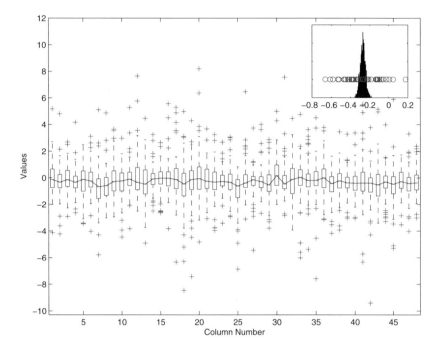

Fig. 6.3   A box and whisker plot showing the behavior of the log ratios (M) for each printing pin separately. There are a total of 48 pins, each one producing 80 spots. Each box has lines for lower quartile, median, and upper quartile values. The whiskers are lines extending from each end of the box to show the extent of the rest of the data. Outliers, which are data points beyond the ends of the whiskers, are shown with + signs. The panel inside the figure shows a bootstrap histogram of the mean of the 48 medians of the log ratios. The circles overlaid on the histogram show the actual values of the 48 medians.

In Figure 6.3, we have plotted a box and whisker plot showing the behavior of the log ratios (M) for each pin separately. There are 48 pins in total, each one producing 80 spots on the particular array that we have used for this example. Each box, corresponding to one of the pins, has lines at the lower quartile, median, and upper quartile values. The whiskers are lines extending from each end of the box to show the extent of the rest of the data. Outliers, which are data points beyond the ends of the whiskers, are shown with + signs. The median values have been connected with a line to improve visualization of the variation between different pins. By visual inspection alone, it might appear as if the medians of the log ratios are fairly consistent throughout the 48 pins. One way to assess whether any given median log ratio is significantly different from the mean of all median log ratios is to apply the bootstrap[4] method to the estimator of the mean of the 48 medians in order to assess its standard error and provide confidence intervals. The panel inside the figure shows a bootstrap-derived histogram of the mean, using 1000 bootstrap samples. The overlaid circles show the actual 48 medians, on the same axis. It can readily be seen that quite a few of these medians are outside the range of the histogram, implying that they are significantly different from the mean of the medians. A bootstrap-based hypothesis test could also be performed, where the confidence intervals can be calculated from the appropriate percentiles of the histogram.

In certain cases, in addition to normalizing the location (mean or median) of the log ratios, it is also advisable to normalize the scale (variance or its robust versions) of the log ratios. For example, for pin-dependent location normalization, the effect would be that all boxes in Figure 6.3 are centered around zero (in the case of median normalization), but their sizes would be different, indicating differences in scale (variability) between the pins. Scale normalization consists of applying a power transformation of the form $X' = X^{1/a}$ to both the red and green channels. The result of such a transformation would be that all boxes in the box and whisker plot would be of approximately the same size, since the inter-quartile range is a measure of scale. There is an inherent assumption that the scale of the log ratios should be approximately the same for all pins. More information about the scaling factors can be found in (Quackenbush, 2002; Yang YH *et al.*, 2002).

6.1.4 Control genes

So far, we have addressed the problem of correcting for systematic sources of variation by applying various transformations to the data, where these transformations are determined from some set of gene expression measurements. For example, in the case of global normalization discussed above, the normalizing constant, such as k_{mean}, is computed using all n genes on the array. As already mentioned, this is reasonable if we assume that the sum of all the measured hybridization intensities should be the same for both samples. This may be the case if most genes do not differ in expression between the two samples and that there is a roughly equal number of differentially overexpressed genes in each sample.

An alternative approach is to use some subset of genes for the normalization. For example, so-called *housekeeping genes*, such as glyceraldehyde-3 phosphate dehydrogenase (GAPDH), β-actin (ACTB), tubulin α1 (TUBA1), and others, are used traditionally as control genes, as they are expected to be

---

[4]The bootstrap is a simulation method used for statistical inference. It can be used to determine the variability of an estimate and provide confidence intervals for parameters when it is difficult or impossible to determine the distribution of the estimator in the usual way (see Efron and Tibshirani, 1993).

expressed ubiquitously at stable levels under different biological conditions. Unfortunately, there is strong evidence that many commonly used house-keeping genes exhibit considerable variability of expression within and across microarray data sets (Lee *et al.*, 2002). In addition, Lee *et al.* (2002) failed to identify mammalian genes that can qualify as 'control genes' in the sense of being stably expressed. Finally, many of the housekeeping genes are expressed at high levels, thus preventing estimation of the dye biases for genes expressed at low levels if the dye bias is intensity dependent. These findings suggest that the use of individual genes for normalization in experiments that compare biological tissues would lead to substantial errors in subsequent analysis of the data.

6.1.5   Dye swap

Our discussion has centered on situations where there is a need to correct for systematic sources of variation within a two-color microarray. In other words, we have attempted to normalize differences between the red and green channels on a per-slide basis, taking into account possible intensity dependence. It is also conceivable that certain individual genes will incorporate one dye more efficiently than the other. If that is the case, the gene expression will be confounded with dye incorporation efficiency, and it may be impossible to tell if a gene is overexpressed in one sample relative to another sample because of a real biological difference or because that gene incorporated the dye more efficiently. A popular approach that can be taken is the *dye swap* method or *paired-slides normalization*. The idea behind this approach is to perform the same two-sample hybridization experiment twice, but with the dyes (Cy3 and Cy5) reversed. Ideally, those genes that are overexpressed in one channel should also be overexpressed in the other channel on the dye-reversed slide, with the same magnitude of the log ratio.

In order to explain how to use this approach to estimate the true log ratios, let us assume that the observed log ratios can be expressed as a sum of the true log ratios, some offset due to dye bias, and a noise term with zero mean. That is, for the kth slide, $k = 1, 2$, the log ratio can be expressed as $M_i^{(k)} = \mu_i^{(k)} + l^{(k)} + \varepsilon_i^{(k)}$, where $\mu_i^{(k)}$ is the true log ratio, $l^{(k)}$ denotes the dye-bias effect, and $\varepsilon_i^{(k)}$ are independent and identically distributed random variables with zero mean, corresponding to the kth slide. Ideally, we would expect that $\mu_i^{(1)} = -\mu_i^{(2)}$; in other words, the log ratios should be of the same magnitude, but with opposite signs. If we assume that $l^{(1)} \approx l^{(2)}$, it follows that

$$\frac{1}{2}(M_i^{(1)} - M_i^{(2)}) = \frac{1}{2}\left(\mu_i^{(1)} - \mu_i^{(2)} + l^{(1)} - l^{(2)} + \varepsilon_i^{(1)} - \varepsilon_i^{(2)}\right)$$
$$= \mu_i^{(1)} + \frac{1}{2}(\varepsilon_i^{(1)} - \varepsilon_i^{(2)}).$$

Therefore, $\frac{1}{2}(M_i^{(1)} - M_i^{(2)})$ is an estimate of the true log ratio $\mu_i^{(1)}$. Yang *et al.* (2001) called this method *self-normalization* because it does not require explicit normalization and only involves combining the observed log ratios in the manner noted above. As Fang *et al.* (2003) writes, "self-normalization is capable of removing dye bias without identifying the nature of that bias."

Dye swap experiments can result in a significant increase in the cost and complexity of experiments, but can potentially improve the quality of the results. For example, Liang *et al.* (2003) recently showed that "dye switching and biological replication improved the reliability of microarray results, with dye switching likely having even greater benefits." Fang *et al.* (2003) report that "employment of dye-flip replicates is critical for obtaining better results through normalization." Workman *et al.* (2002) notes that "dye-swap normalization was less effective at Cy3-Cy5 normalization than [intensity-dependent] methods alone." Dobbin *et al.* (2003) show that "reverse labelling of individual arrays is generally not required" and that "it is not necessary to run every sample pair twice so as to eliminate the dye bias." Thus, it is important to carefully evaluate whether the improvement in reliability merits the additional investment of time and resources, for the investigator's particular problem setting and data set quality.

6.1.6   Single-channel normalization for multiple arrays

In many instances, one may be interested in obtaining data from a number of microarrays, but not in the form of ratios. For example, we might want to analyze 30 different tumor specimens using unsupervised methods, such as multidimensional scaling, to determine how the tumors cluster together and form subgroups. Or we may be interested in assessing the intratumor heterogeneity by arraying a number of different sections of a tumor and comparing the gene expression profiles. In order to render the analysis meaningful, it is important to ensure uniformity and comparability between the arrays. So why not use ratios anyway? We could, after all, cohybridize the tumor samples with some reference sample, such as a mixture of cell lines or tissues, that is labeled with the same dye on all the arrays and simply use the (properly normalized) ratios for our subsequent analysis. Presumably, the rationale for using ratios in this manner is to furnish built-in normalization: If there are differences between spots, such as artifacts due to dust or scratches, or regional differences within the array, both channels will reflect these variations and when the measured intensity in one channel is divided by the measured intensity in the other channel, these differences will cancel out. However, this approach is valid if each spot in the reference channel is expressed at some reasonable level such that when its intensity appears in the denominator, the ratio will not be unduly large. However, even if a mixture of different cell lines is used as a reference, there will still be many genes that are expressed at levels comparable to background (recall that the distribution of intensities is lognormal, meaning that most genes are expressed at low levels) and hence, after background subtraction, will result in zero or close-to-zero intensities. Consequently, dividing by these numbers will become meaningless and the corresponding genes in the tumor sample will be rendered uninformative, as there would be no way of relating the ratio to the absolute expression level of the gene in the sample. An alternative strategy, should one still prefer to use ratios, is to create a reference that will express ubiquitously throughout the array, for example by using oligos complementary to a common sequence present in all microarray features (Hu *et al.*, 2002; Dudley *et al.*, 2002).

Let us now return to the problem of normalizing a number of single-channel microarray measurements. The underlying issues are the same as those we described in previous sections. However, instead of comparing two samples, we are now comparing k samples on k microarrays, all labeled with the same dye.[5] Perhaps the simplest approach is to apply global normalization to all the arrays by dividing all expression measurements in an array by the median (or mean, trimmed mean, percentile, etc.) of that array. Of course, control genes could also be used, but this is not recommended for reasons already stated. In order to incorporate intensity dependence, we could modify the lowess-based normalization. Let $M_i^{(p,q)}$ and $A_i^{(p,q)}$ be the log ratio and mean log intensity, respectively, of the ith gene ($1 \leq i \leq n$) between the pth and qth array ($1 \leq p, q \leq k$). As an example, consider Figure 6.4, where we have plotted $M_i^{(p,q)}$ vs. $A_i^{(p,q)}$ using the Cy5 channel of 4 cDNA arrays. As can be seen from the fitted lowess curves, there is a need for intensity-dependent normalization.

One approach that can be taken is to apply the lowess-based normalization to each pair of arrays [Bolstad *et al.*, (2003) called this method *cyclic lowess*]. In general, there are $k(k-1)/2$ normalizations to be performed. For each pair (p, q), let $l^{(p,q)}$ be the normalization function and $X_i^{(p)}$ be the log intensity of the ith gene on the pth array ($1 \leq p \leq k$). The normalization consists of the transformations $X_i'^{(p)} = X_i^{(p)} - l^{(p,q)}/2$ and $X_i'^{(q)} = X_i^{(q)} + l^{(p,q)}/2$. Then, the normalized log ratios become

$$
\begin{aligned}
M_i'^{(p,q)} &= X_i'^{(p)} - X_i'^{(q)} = X_i^{(p)} - l^{(p,q)}/2 - (X_i^{(q)} + l^{(p,q)}/2) \\
&= X_i^{(p)} - X_i^{(q)} - l^{(p,q)} = M_i^{(p,q)} - l^{(p,q)},
\end{aligned}
$$

which is the same as for two-channel dye-bias normalization. Similarly,

$$
\begin{aligned}
A_i'^{(p,q)} &= \frac{1}{2}(X_i'^{(p)} + X_i'^{(q)}) = \frac{1}{2}(X_i^{(p)} - l^{(p,q)}/2 + (X_i^{(q)} + l^{(p,q)}/2)) \\
&= \frac{1}{2}(X_i^{(p)} + X_i^{(q)}) = A_i^{(p,q)}.
\end{aligned}
$$

---

[5] The material in this section can also apply to radioactively labeled nylon membrane microarrays, which are intrinsically single-channel.

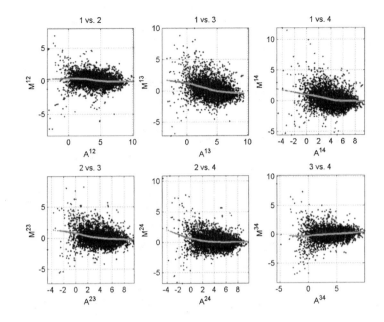

Fig. 6.4   The six scatter plots show all combinations of $M_i^{(p,q)}$ vs. $A_i^{(p,q)}$ using the Cy5 channel of four different cDNA microarrays. The fitted lowess curves clearly indicate a need for intensity-dependent normalization. See insert for a color representation of this figure.

After applying this normalization to all pairs of arrays, the method is iterated. Bolstad *et al.* (2003) found that one or two complete iterations are usually sufficient.

6.2   REPLICATION

Replication is a basic principle in experimental design. It involves making independent observations under the same experimental conditions and carries tremendous implications for quality control. This issue is particularly relevant in the design of microarray experiments, where we can distinguish two different kinds of replication: intra-array and inter-array replication. Intra-array replication refers to measuring the same gene via several different spots on the same microarray. Inter-array replication refers to repeating the same hybridization experiment on several different microarrays. It is obvious that both inter- and intra-array replication can produce more consistent and reliable findings and increase the overall quality of the data analysis at the

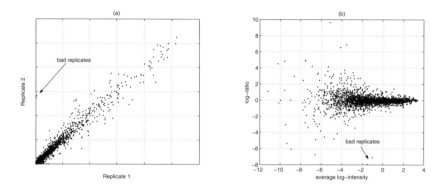

Fig. 6.5 (a) A scatter plot of the measured intensities of two replicates of each gene. (b) The same data plotted as log ratios vs. mean log intensities of each of the two replicates. Two obvious bad replicates are indicated by an arrow in both scatter plots.

expense of increased cost and amount of biological material used. Let us first consider intra-array replication with an example that illustrates its usefulness.

6.2.1   Intra-array replication

As our example, we will use a cDNA microarray with 4800 spots and will focus for the moment on only one channel. This microarray contains 2303 genes, each printed in duplicate for a total of 4606 spots. The duplicate spots are printed in different regions of the array to rule out the possibility that an artifact will affect both spots. The other 194 spots, arranged throughout the array, are control spots containing housekeeping genes as well as positive and negative controls. Figure 6.5(a) illustrates a scatter plot of the measured intensities of the two replicates for each gene. Ideally, all points on the scatter plot should lie on the diagonal, meaning that both replicates should be equal. We can notice at least two obvious bad replicates, indicated by an arrow pointing to them. We can also visualize the concordance between the two replicates using a plot that is similar to the M vs. A plot, where we plot the log ratio of the two replicates vs. their average log intensity, as shown in Figure 6.5(b). The same two bad replicates are also indicated by an arrow. Note that for very low average log-intensities, meaning that both replicates are very small, the log ratios can be much more variable, as expected, and may not constitute bad replicates. This fact suggests that the threshold for automatically determining whether two replicates are 'bad' should depend on their average intensity.

There are a number of ways that one could define such thresholds. For example, Tseng *et al.* (2001) suggest using the coefficient of variation[6] (CV) as follows. First, the CV is plotted against average intensities of the replicates for each gene. For each gene, the 50 genes whose average intensities are closest to the given gene are used to compute the percentile of the CV of that gene. If that gene is in the top 10% among the genes in its windowing subset, it is deemed unreliable. Another somewhat simpler and computationally less intensive approach might be to fit a smooth curve (again by lowess or some other method). For example, one could fit a curve to all positive log ratios and another curve to all negative log ratios and set the thresholds (at both ends) to be equal to some constant times the respective curves. This idea is illustrated in Figure 6.6, where each lowess curve is multiplied by 4. Yet another approach, proposed by Baggerly *et al.* (2001), might be to use a locally smoothed estimate based on the interquartile range (also a measure of spread), using some fixed window width, and plotting the threshold curves as some constant (e.g., 3) times this interquartile range.

We would like to emphasize two important aspects of any approach that uses some thresholding process to demarcate unreliable replicates. First, a threshold, by its very nature, dichotomizes the data—in our case, into reliable and unreliable replicates. However, it is up to us to decide what action to take with respect to data that is deemed unreliable. We can either perform the analysis without it, replace it with other data that we can estimate, or choose to use it, possibly giving it less weight, the latter depending on some quality metric that we can define. In the context of replication, some measure of spread will serve this purpose. Thus, what is important in this process is not that some replicates are marked as unreliable, but what we choose to do with these replicates. Second, most, if not all methods used to define these thresholds are heuristic in nature. Indeed, in the examples above, the 50 genes with closest intensities and the top 10% of the coefficients of variation are quantities that are certainly chosen heuristically. Similarly, the lowess curve or the smoothed interquartile range, as well as their multiplicative factors (4, 3, etc.) are also heuristic. Of course, certain modeling assumptions can be made, such as the underlying distribution of the data, in order to justify one method over another: A certain multiplicative factor multiplied by some measure of spread estimates a certain number of standard deviations under some assumed distribution. But ultimately, even the number of standard deviations that we choose to determine an unreliable replicate is still chosen heuristically. This aspect should not be viewed in a negative fashion, since many of these methods can be quite useful in dealing with unreliable replicates and preventing them from confounding subsequent analysis. The decision about which method to use should be made in view of the goals of the analysis and the particular

---

[6] The coefficient of variation is defined as the standard deviation divided by the mean. It can be used as a measure of spread for a set of data.

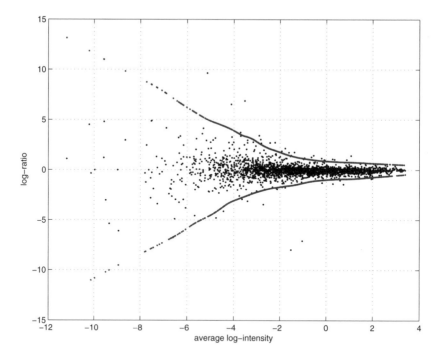

Fig. 6.6 A possible approach to detect unreliable replicates. Two loess curves are fitted to the positive and negative log-ratio parts of the scatter plot and then multiplied by 4. All replicates that our outside these curves are marked as being unreliable.

aspects of the data set being analyzed. As a final recommendation, we strongly encourage the investigator to visually assess the data represented in some graphical form. For example, simply viewing the log ratio vs. mean log intensity scatter plot can reveal quite much about the overall data fidelity in terms of replicate concordance.

As we have mentioned, intra-array replication is an important aspect of quality control since it can provide a more accurate estimation of the inherent variability in a microarray experiment and can also increase the probability of detecting differentially expressed genes, given that variability. Therefore, an important question arises: How many intra-array replicates do we need? To answer this question properly, it should be framed as a power study, using an estimate of the sampling variation to calculate the probability of detecting significant changes under different sample sizes, as shown recently by Black and Doerge (2002). The approach described uses a so-called *control array*, where a sample is cohybridized with itself using two different dyes, in order to obtain information about the sampling variation.[7] The data are used in conjunction with ANOVA models in order to calculate the number of replicate spots necessary for detecting significant changes in expression with high probability. That is, the residuals from a fitted ANOVA model (Kerr *et al.*, 2000) can be used for power calculations, either via normality assumptions or the bootstrap. Black and Doerge (2002) found that for an *E.coli* data set (Richmond *et al.*, 1999), at least two replicates are necessary for a reasonably high probability of detecting a threefold change in expression, while three replicates were necessary to ensure a high probability of detecting a twofold change. On the other hand, for a yeast sporulation data set (Chu *et al.*, 1998), two replicates were required to detect a twofold change with high probability under the normal-theory model, while the bootstrap method suggested that only one replicate was sufficient to detect a twofold change.

We can make several conclusions. First, the results of power calculations will depend on the particular data set, and no "gold standard" for the minimum number of replicates can be given. In particular, the quality of the entire microarray experiment (image quality, hybridization quality, etc.) will affect the results of such calculations. Second, the results of these calculations will also depend on the modeling assumptions and methods (i.e., the form of the additive model, normality assumptions or nonparametric methods, the type of data preprocessing, etc.). These conclusions suggest that the investigator wishing to know the minimum number of intra-array replicates should first decide the minimum level of fold change required for detection with a high probability (the latter also decided *a priori*) and make this determination using a power study framework for the particular microarray technology and protocols being used.

6.2.1.1 Composite microarrays
In addition to detecting differential expression, microarrays are beginning to be used for quantitative model development. Examples include construction of genetic regulatory networks from gene expression data, design of classifiers from microarray data for discriminating various tissues or cancers, and unsupervised learning or clustering analysis. However, in many instances, microarrays do not possess sufficient sensitivity and reproducibility that are essential for fine-scale quantitative methods. Although intra-array replication is crucial for making reliable inferences and carrying out quantitative analysis with microarray data, there are a number of limiting factors, especially for in-house spotting facilities.

Spotting robots typically have a limitation on the number of spots that can be printed reliably. Thus, increasing the number of replicates can be done at the expense of decreasing the number of genes surveyed. In addition,

---

[7]Although the Cy3 and Cy5 intensities for each gene can be considered as replicates in a control array, it is assumed that systematic sources of variation such as dye bias have been accounted for by normalization.

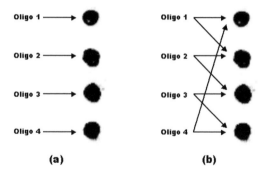

(a)   **(b)**

Fig. 6.7 An illustration of the difference between the classical and composite microarray designs. The drawing in (a) shows that each gene represented in an oligonucleotide (oligo) is placed into its own individual spot on the glass slide. Drawing (b) shows an example where each spot contains a mixture of two different oligos. Thus, it is expected that the measured signal intensity of such a spot would be a combination of the intensities of the constituent genes measured individually.

even if the total number of spots was not a limitation, having more spots requires more labor during the image analysis stage, as most microarray image analysis tools are not totally unsupervised or automatic, which translates directly into higher cost or lower throughput. Finally, fewer spots require less physical space on the solid support (e.g., glass slide), which in turn translates into smaller amounts of RNA required for hybridization. Recently, a novel approach to design so-called composite microarrays was proposed by Shmulevich *et al.* (2003), allowing one to increase the representation of each gene on the array without increasing the number of printed spots by the same factor. A very similar approach, appearing at roughly the same time, was proposed by Khan *et al.* (2003), although their main motivation was to deal efficiently with spot *dropouts*, which are corrupted or missing spots.

The main idea behind this method is illustrated in Figure 6.7. Composite microarrays differ from the usual microarrays in that instead of the 'one gene – one spot' design, several different genes are represented in each spot. This is accomplished by mixing the oligonucleotides corresponding to different genes and spotting the mixtures.[8] The gene expression levels of the individual genes can be recovered from the measured mixture intensities by means of mathematical and statistical methods. This approach is akin to the common engineering practice of combining a number of different signals into each sensor and then recovering the individual signals from the received signal mixtures.

[8] Composite microarrays use spotted long oligonucleotides, rather than cDNA, since high-specificity sequences can be selected to avoid cross-hybridization between the mixed genes in a single spot.

The advantage of such an approach is that each gene appears in a number of different spots located on the microarray (each time with different partners) and thus furnishes a type of replication without increasing the number of printed spots by a factor that would be incurred with the usual intra-array replication, where each spot contains only one gene. As discussed in Khan *et al.* (2003), this also makes it less likely that a particular gene will become uninformative because of a corrupted or missing spot.

6.2.2   Inter-array replication

Inter-array replication refers to repeating a microarray experiment more than once. However, we must be precise about what we mean by a replicated microarray experiment. Suppose that we are working with some cell lines and wish to perform microarray experiments under certain conditions. In order to produce replicated measurements, we could extract RNA from several different cell lines, cultured under as nearly identical conditions as possible, and perform microarray hybridizations using each of those different RNA samples. We could, on the other hand, extract RNA from one cell line, divide it into several parts, and perform hybridizations with each part. In the former approach, we will have to deal with additional experimental variability due to differences in cell lines and their respective RNA extraction steps. However, it can give us insight about the inherent biological variability between the different cell line populations. The latter approach, on the other hand, is more informative about the particular cell line being used, but cannot provide any knowledge about the population differences. As noted by Kerr *et al.* (2002b), it is important to consider the goals of the microarray experiment carefully in order to determine what kind of replication is required for making desired inferences.

Repeated experiments of the first type, where one performs several different RNA extractions, can also be used to assess biological heterogeneity. For example, consider a situation in which we wish to evaluate the heterogeneity of a tumor. This is a very important issue in the context of microarray-based cancer diagnosis, since if the gene expression profiles of samples taken from different locations in the tumors are drastically different, biological classification on the basis of random sample analysis may not be adequate. For example, as part of our study of leiomyosarcoma, Shmulevich *et al.* (2002b) assessed intratumor heterogeneity by analyzing gene expression profiles from several carefully mapped peripheral and core specimens from the same tumor. At the same time, we also performed several experiments in which a tumor sample from each of several patients was hybridized to three arrays using the same RNA (the second type of replication mentioned above). Performing both types of replication allowed us to compare the intratumor variability with the typical variability observed for replicates using the same RNA. That is, if the variability exhibited between the different tumor sections would prove to be considerably higher than that between same-RNA replicates of a single tissue,

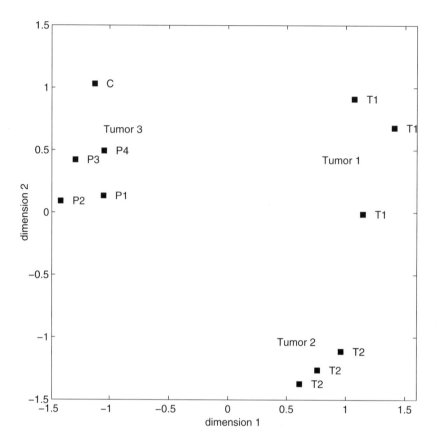

Fig. 6.8 A two-dimensional multidimensional scaling (MDS) solution reproducing proximities between replicate same-RNA samples from patient 1 (T1) and patient 2 (T2), and the peripheral/core sections of a tumor from patient 3 (P1, P2, P3, P4, and C, respectively).

this would imply that factors other than microarray experimental conditions led to the variation observed.

A good way to assess such variability or heterogeneity visually is by means of multidimensional scaling (MDS), which is a class of methods used for constructing a low-dimensional representation from a matrix of similarities or dissimilarities derived from multidimensional data. In other words, MDS attempts to model the dissimilarities in the multidimensional space as distances between points in a geometric space. A good reference is (Borg and Groenen, 1997). Figure 6.8 illustrates a two-dimensional MDS solution derived from a matrix of Euclidean distances between the profiles of normalized log ratios corresponding to each of the following samples: three same-RNA replicates of tumor 1 (patient 1), three same-RNA replicates of tumor 2 (patient 2), and four peripheral and one core section of tumor 3 (patient 3), all cohybridized to a normal adjacent tissue from patient 3 used as a reference. An informal inspection of Figure 6.8 suggests that the variability exhibited by different tumor sections was well within the observed experimental variability of same-RNA replicate experiments.

Same-RNA inter-array replication can also be very useful for estimating reproducibility. If repeated measurements exhibit high variability (e.g., variance), it would imply that reproducibility is poor. A good quality control measure to assess reproducibility is the coefficient of variation (CV), which we discussed in Section 6.2.1. The CV reflects the fact that the variance of repeated measurements depends on their mean intensity: Highly expressed genes are more variable. This is illustrated in Figure 6.9(a), where we have plotted the standard deviations vs. the means for each gene across eight replicate microarrays, each one hybridized with the same mixture of 10 different cell lines. Only the green-channel measurements are used. It can clearly be seen that the standard deviation of the replicated measurements increases with mean intensity. The scatter plot was truncated at mean intensity of 500 in order to show the linear portion of the dependence of standard deviation on the mean. For even larger intensities, the slope begins to decrease somewhat, losing its linearity. Figure 6.9(b) shows the CV plotted against the mean intensity. For very low intensities, the CV itself is highly variable; for most other intensities, it is approximately constant. As already mentioned, for very high intensities, the CV decreases somewhat. The constant-CV region can be useful as a measure of reproducibility. For example, if one decides to change some aspect of the microarray production protocol, one can assess the effect of this change simply by producing a number of replicate arrays, as shown in this example, and seeing whether the CV is increased or decreased relative to what it was before the change in the protocol. A decrease in the CV would indicate an improvement in quality.

6.3 MISSING VALUES

Recall that intra-array replicates can be used to flag unreliable data in a microarray experiment. That is, if the two replicate measurements are overly discordant, either in terms of intensities (single channel) or log ratios, as compared to most other differences between replicates with the same intensity, we can choose to flag the corresponding gene as being unreliable and decide later how to use this information. If the replicates are concordant and we decide to label the gene as reliable, we can simply combine the values of the replicates, for example by averaging, to form a single estimate of the gene expression or log ratio. However, when no estimate can be formed from the replicates, we will have to handle the resultant missing values if we intend to use the microarray data for subsequent analysis.

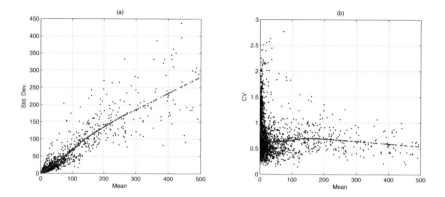

Fig. 6.9 (a) A scatter plot showing the standard deviations vs. the means across eight replicate microarrays, each one hybridized with the same mixture of 10 different cell lines. Only the green channel is shown. The scatter plot was truncated at a mean intensity of 500, to show the linear range (constant CV); (b) the CV plotted against the mean. It can be seen that for most of the range, except for very low intensity, it is approximately constant, with a mean CV equal to 0.75. In both scatter plots, a lowess smoothing curve was fitted to show the trend.

Missing values can also occur for other reasons, especially in the case of microarrays with no replicated spots. For example, some spots might become corrupted due to a scratch, dirt, or some other artifact. Consequently, the image analysis software, using some quality metric, will not be able to produce a reliable estimate of the intensity, resulting in missing data for the gene represented in that spot. As another example, some robotic spotters, such as those with split-pin technology, can have spot dropouts either due to bent or clogged pins or insufficient uptake of DNA solution for printing. This is discussed in more detail in Chapter 2.

We have several options as to how to deal with missing values. The easiest approach is simply to get rid of any genes that have a missing value in any of the samples. In other words, if the data are summarized in a data matrix, where each row corresponds to a gene and each column corresponds to a sample (microarray experiment), then all rows that contain a missing value would be removed. This approach, although simple, may be useful when gene expression data are used in a global fashion, such as in multidimensional scaling. In other words, if we are merely interested in visualizing proximities between samples (or clustering them), or if we are comparing entire distributions of spot intensities (e.g., Hoyle *et al.*, 2002), removing a small percentage of genes may not be unduly detrimental to the analysis. Nonetheless, there are other approaches that can be taken (Troyanskaya *et al.*, 2001).

For example, another simple approach is to replace each missing value with the average of all the other values of that particular gene, that is, by the average of the entire row, ignoring any other missing values in that row. Although this method is simple to implement, it does not take into account any correlation structure in the data, and another method, based on k-nearest neighbor (kNN) analysis, was proposed by Troyanskaya *et al.* (2001). The basic idea behind the kNN method is to select k other genes with the gene expression profiles (i.e., rows) most similar to the gene containing the missing value to be estimated. Similarity is measured by Euclidean distance between the gene profiles. The actual estimate of the missing value is a weighted average of the values of these other k selected genes in the same experiment (microarray), where the weight of each gene is determined by the similarity of its overall profile. This method was found by Troyanskaya *et al.* (2001) to be quite accurate for several publicly available data sets. Also, the method was shown to be relatively insensitive to the exact value of k within the range 10–20 neighbors.

6.4 DATA MANAGEMENT ISSUES

In many cases, factors other than measurement noise, experimental variability, or systematic sources of variation can have a considerable effect on the quality of the data and the ensuing conclusions based on the analysis of those data. These factors often originate from poor data management procedures and/or inadequate communication between group members. Let us give a simple example of a possible pitfall to illustrate this point.

Data from image analysis software is typically summarized in a text file or spreadsheet in which every row contains a number of measurements of a particular spot. The columns contain spot volume, position, area, standard deviation, background, and other quantities of interest. The first column of such a data file often contains spot labels, which often code for the position on the array. For example, the software *ArrayVision* by Imaging Research, Inc. can output spot labels such as "A - 1 : B - 10" or "C - 12 : H - 9." In this case, the first letter–number combination codes for the subgrid (pin) on the array, while the second letter–number combination codes for the position of the spot in that match. This coding system can be very useful for facilitating visual inspection of the microarray images—one can look in the data file, read the intensity of a certain spot, and quickly find it on the microarray image.

The potential pitfall may take the form of something as seemingly innocent as the sorting of the rows in the spreadsheet. Let's suppose that lab A produces arrays with two intra-array replicates per gene and uses some image analysis software, such as *ArrayVision,* to produce the data files. Let us also suppose that a batch of the same arrays is sent to another lab (B), which performs its own hybridization, scanning, image analysis, and eventual data generation. After this, B sends back all the data to A for statistical analysis. However, B's software (or spreadsheet) is configured differently, such that it sorts the rows alphanumerically based on the spot labels. An example is

Spot labels	VOL - Levels x mm2	SD - Levels	Pos X - mm	Pos Y - mm	Area - mm2	Bkgd	sVOL	S/N
A - 1 : A - 1	79.5464	715.41	4.383	20.395	0.028	48.372	31.174	6.253
A - 1 : A - 10	62.7444	368.6	8.026	20.375	0.028	49.008	13.736	2.68
A - 1 : A - 2	49.1416	181.38	4.784	20.395	0.028	48.428	0.714	0.141
A - 1 : A - 3	57.789	238.7	5.204	20.395	0.028	48.898	8.891	1.709
A - 1 : A - 4	53.3207	231.95	5.584	20.395	0.028	48.732	4.589	0.896
A - 1 : A - 5	61.4809	364.65	6.005	20.375	0.028	49.727	11.754	2.228
A - 1 : A - 6	65.9933	446.58	6.405	20.415	0.028	49.644	16.349	3.065
A - 1 : A - 7	55.4423	294.97	6.785	20.375	0.028	48.704	6.738	1.24
A - 1 : A - 8	57.1392	314.86	7.185	20.375	0.028	49.036	8.103	1.385
A - 1 : A - 9	82.6791	548.62	7.606	20.375	0.028	49.34	33.339	6.075
A - 1 : B - 1	53.0883	189.73	4.383	19.995	0.028	48.538	4.55	0.839
A - 1 : B - 10	59.0421	312.73	7.986	19.955	0.028	48.981	10.061	1.819

Fig. 6.10   A small part of a data file produced by *ArrayVision*. Note that the spot labels in the first column are sorted alphanumerically (in a spreadsheet), along with their corresponding rows.

shown in Figure 6.10. Note how the second row is "A - 1 : A - 10" instead of "A - 1 : A - 2," as one would expect.

Such a simple difference can have serious repercussions. For example, if one would proceed to perform any kind of statistical analysis with the data set sorted in the wrong manner, genes that are expected to be differentially expressed might not be, and vice versa; clustering analysis would probably yield meaningless results; and supervised learning or classification algorithms would produce the wrong gene names that discriminate the class labels. The reason behind this is that lab A expects a certain gene to be associated with spot "A - 1 : A - 2," although in reality it is the gene that is associated with spot "A - 1 : A - 10" that is in its place.

Problems such as this can easily be caught and prevented simply by viewing the data in some suitable form. For example, if the array has two replicates for each gene, the statistical analysis software that reads the data files would expect to see the replicates in certain positions (rows) of the data file. However, because of different sorting, the replicates would be in entirely different positions. Consequently, the data would appear to be highly unreliable if viewed on a simple scatter plot. Figure 6.11(a) shows a scatter plot of the two replicates under normal sorting, while Figure 6.11(b) shows a scatter plot obtained from an incorrectly sorted data file. Clearly, the second scatter plot reveals a highly suspicious situation where no concordance between replicate spots is evident. For arrays that have no replicate spots, a procedure for viewing the expression levels of certain control genes, such as negative or positive controls, would also reveal an abnormality if the rows of the data file are sorted incorrectly. One would expect that negative controls (blank spots or probes from different organisms that are not expected to hybridize to the sample) would exhibit a consistently low expression relative to positive controls (e.g., housekeeping genes).

In addition to row sorting, a related problem can take the form of reading the wrong column of the data file. Most image analysis software programs

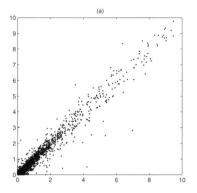 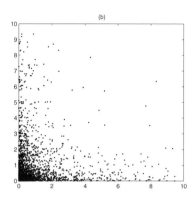

Fig. 6.11   (a) A scatter plot showing raw intensity values of two replicate spots for each gene. As this is based on correct sorting of the data file, a high degree of concordance can be seen. (b) The scatter plot obtained from an incorrectly sorted data file, showing a very low degree of concordance between replicates and indicating a problem.

allow the user to customize which quantities corresponding to each spot will be output and in which order. For example, some users may choose to output the horizontal and vertical positions of the spots, the spot area, the background levels, and so on, while another user may choose not to output the spot area, which would result in a data file with that column absent. Consequently, if lab A expects the background-subtracted spot volume to be in column 7 and lab B has it in column 6, lab A's statistical analysis software will be reading the wrong column. Once again, such simple problems can be prevented by looking at the data in some form. Consider, for example, the distribution (histogram) of the data. We know that gene expression data should roughly follow a lognormal distribution, which would look like the histogram shown in Figure 6.12(a). That is, most genes would be expressed at very low levels (close to zero for background-subtracted data) and very few genes would be expressed at high levels. Figure 6.12(b), on the other hand, shows the histogram of the spot background intensities. This histogram looks roughly like a normal distribution, and its range of values is very small (between 47 and 53). A histogram of this shape should immediately signal a problem.

There are, of course, many other possible scenarios that could lead to problems similar to those that we described above. Generally speaking, various miscommunications or data-handling inconsistencies can potentially occur in a large group or several groups working together and sharing data. We can, therefore, make a general recommendation: Always apply sanity checks to your data before proceeding to analyze them. The best way to screen for problems is to look at the data. Scatter plots of replicates should have a certain shape. So should M vs. A scatter plots. Distributions of intensity data should also be fairly consistent. If the data look unusual, such as in Figures

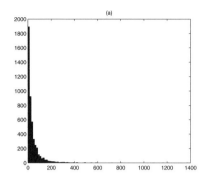 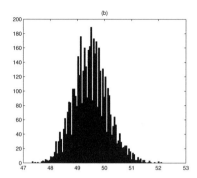

Fig. 6.12 (a) The histogram of background-subtracted raw intensity data, roughly following a lognormal distribution. (b) The histogram of background intensities, which looks more like a normal distribution.

6.11(b) and 6.12(b), it probably means that there is a problem. *Caveat lector*: If the data appear normal, it does not necessarily imply that nothing is wrong.

6.4.1   MIAME: necessary information to facilitate data sharing

Although the aforementioned data management issues are very much a reality in a typical team working with microarray data, the problems can be compounded manyfold when microarray data are publicly released and other teams are in a position to analyze these data and try to make conclusions from them. For this reason, there has been an enormous push in the community to establish guidelines to facilitate the interpretation and verification of microarray results. This has culminated in the MIAME (Minimum Information About a Microarray Experiment) standard (Brazma *et al.*, 2001), which is now required by a number of journals, such as *Nature, Cell, Lancet,* and others. The MIAME standard requires information on experimental design, sample preparation and labeling, hybridization procedures and parameters, measurement data and specifications, and array design. The experimental design contains information about the type of experiment, such as normal vs. tumor or time-course data, the experimental factors, detailing the tested conditions, such as duration or dose of treatment, number of hybridizations, the type of reference used, and the type of quality control steps taken, such as replicates and dye swaps. The information about the samples that are used should consist of their biological origin, their characteristics, protocols for their manipulation (e.g., growth conditions, treatments, etc.) and preparation for hybridization, as well as labeling protocols. Hybridization procedures, conditions, and protocols should also be described. Important parts of MIAME are the actual measurement data and description of specifications, which should include the quantitations based on the images and those quantitations on which the authors based their conclusions. Also important is the inclusion of information about scanning hardware and software, image analysis software, description of the measurements that are output by the image analysis software, and data transformation procedures. Finally, a section on array design should include the type of array used (e.g., glass, synthesized), surface and coating specifications, location and ID of each feature on the array, the source of the reporter molecules (e.g., cDNA or oligo collection source), spotting protocols, and any additional treatments performed prior to hybridization. Although details of different experiments may be different, MIAME aims to define the core of any microarray experiment. As such, it is not a formal specification but a set of guidelines. The Microarray Gene Expression Data Society (MGED, http://www.mged.org/), which developed MIAME, has made available a MIAME checklist that can be very helpful for presenting microarray results according to the MIAME standard. It is recommended that users who exchange microarray data follow these guidelines such that others can assess the data quality and verify the results and conclusions independently.

6.5 SMALL-SAMPLE-SIZE ISSUES

Less is More.

—Ludwig Mies van der Rohe (Architect; 1886–1969)

What is *small*? This question may leap to mind and begs to be asked when one reads the title of this section. Is there some magic sample size that is necessary in order for us not to regard it as being small? If not, then there must be a continuum in accordance with which the results of statistical analysis methods are somehow dependent on the sample size. If so, then this commonly used terminology—small sample size—does not bear a precise mathematical definition but merely reflects a certain state of affairs whereby certain assumptions of the employed statistical methods may be violated or our confidence in the results of certain methods may be very small. Thus, this term is intimately connected with the particular method that is applied.

Microarray data are often used for studying differences on the molecular level between various samples, such as normal and diseased tissues. The most common scenario is detection of differential expression of genes between the two sets of samples. In other words, for every gene we form some test statistic and test the hypothesis that there is no difference in expression between the two sets of samples using, say, a t-test or Wilcoxon test. We can either have a number of paired samples, using the two-channel array format, or a number of single-channel intensity measurements from each set of samples. It is quite clear that the larger the sample size (number of replicates), the more power[9] the test will have under a given type I error rate. A sample size/power calculation can be made in order to determine the sample size required to correctly detect a certain level of differential expression with high probability (Black and Doerge, 2002; Pan *et al.*, 2002). We shall not concern ourselves here with this problem and refer the reader to a number of excellent recent papers on various approaches for detecting differential expression (Pan, 2002, 2003; Tsai *et al.*, 2003; Dudoit *et al.*, 2002; Troyanskaya *et al.*, 2002; Huber *et al.*, 2002; Tusher *et al.*, 2001; Tsodikov *et al.*, 2002; Chen *et al.*, 2002; Yang IV *et al.*, 2002; Pan *et al.*, 2002).

In this section we take a different viewpoint: Given that we have a sample of a certain size, what are some of the quality control issues as regards the design, performance, and analysis of statistical methods used to analyze these data? Let us focus on the important problem of gene expression-based classification, which has become a popular tool in computational genomics and microarray communities. The goal of classification is to take a gene expression profile (i.e., microarray measurement of some sample) and predict the class that contains this sample. The class can be *normal* or *tumor* or perhaps *sensitive* or *resistant* to some treatment. In general, there can be many classes. The classifier is designed from the gene expression data. The design itself can be thought of conceptually as consisting of two stages. The first stage determines which genes (features) should be used as classifier variables (i.e., inputs to the classifier). This stage is often called *feature subset selection*. The second stage actually constructs the classifier itself, using some rule, from the gene expression data obtained from the selected feature genes. Since expression data exhibit randomness on account of biological and experimental variability, the design and performance evaluation of the classifier must take this randomness into account. In his excellent review paper on this topic, Dougherty (2001) points out three critical issues arising in microarray-based classification, which we describe below.

6.5.1   Classifier design

The first issue is concerned with what is referred to as model selection in statistics. A fundamental question is: What type of classifier should we choose? In the field of machine learning, the model (i.e., the type of classifier or rule) is called the *hypothesis space*. Once we determine the type of classifier, meaning a class of classifiers, the next step, of actually producing a particular classifier from this class (i.e., selecting a particular hypothesis from the hypothesis space), is fairly well understood, as it typically involves the estimation of parameters. The classifier design issue is critical because our ultimate aim

---

[9] The power of a test is the probability of rejecting the null hypothesis when it is false. In the context of microarrays, this is essentially the probability of correctly detecting a change in expression.

should be to provide good classification performance over the general population from which our sample is taken. That is, the classifier should exhibit good predictive accuracy.

The fundamental principle in model selection is that a complex model may be able to describe the observed data very well, but it may be very poor in predicting future data. If the data contain random fluctuations or noise, an excessively complex model will *overfit* the data along with the noise and will obviously provide a poor fit to future data. The goal in model selection (in our case, classifier design) is to strike the right balance between simplicity of the model (classifier) and its goodness of fit. It would be unwise to design an overly complex classifier, say with hundreds or thousands of parameters to be estimated, when the number of samples is small (typically, less than 100). Such a classifier is likely to exhibit a very small error on the available data set but may have very high error on unseen data. Using appropriate criteria or methods to strike this balance will result in the best predictive accuracy, which should be our ultimate goal. The Bayesian information criterion, Akaike's information criterion, minimal description length principle, normalized maximum likelihood, and cross-validation methods are but some of the methods that can be used for this purpose. In short, one must always heed William of Ockham's (d. 1349) famous maxim: *Plurality should not be assumed without necessity.*

6.5.2   Error estimation

The second issue is related with the estimation of the error produced by the designed classifier, especially in the context of limited data. After we have constructed a particular classifier from the data (i.e., selected a particular hypothesis from the hypothesis space), we must be able to evaluate its performance in terms of its predictive accuracy. That is, we must estimate the error that the designed classifier makes over the population. This is necessary in order for us to compare classifiers and choose the one with the best predictive accuracy.

Since we only have the availability of a certain amount of data from the population, one common approach is to split these data into *training* and *test sets*. The classifier is then designed from the training data set (e.g., two-thirds of the data), and the error is estimated from the proportion of errors it makes on the test data set (e.g., one-third of the data). In the presence of large sample sizes, this approach is appropriate, with the estimates of the error being quite reliable. The underlying assumption is that the training and test sets are representative of the population.

For the case of limited data, neither the training nor test sets are large enough and the error estimates begin to have a high variance, leading to unreliable estimation of classifier performance. The problem is that many error estimates can be very low, due to their high variance, giving us an unrealistically optimistic picture of the designed classifier. A common small-sample approach is *k-fold cross-validation,* whereby the data set is randomly split into k subsets of equal size, and each subset is used for testing and the remainder for training.[10] The estimated errors from all the test sets are averaged to obtain an overall error estimate. A special case of cross-validation is *leave-one-out* estimation, which is just n-fold cross-validation. Thus, exactly n classifiers are designed, where n is the sample size, from $(n - 1)$-element training sets and tested on the remaining 1-element test sets.

There are many nuances in error estimation, such as how to select the appropriate proportion of the data for splitting into training and test sets and whether or not to stratify the data (ensuring that each class is represented with approximately equal proportions in the training and test sets). All these issues are important for obtaining good-quality estimates of the classification error. An important rule in error estimation that is sometimes neglected is that the test-set data should never be used for anything during the training step. Not heeding this rule will lead to inaccurate error estimates and will negatively affect feature subset selection, which we discuss next.

6.5.3   Feature subset selection

Not all of the features (genes) are useful for learning the classification rule. Some features are irrelevant for discrimination between the classes, and some are redundant with other features. Although it may seem that such unnecessary features should not degrade the performance of a classifier, in practice, they often do. Thus, the aim of feature subset selection is to choose a subset of genes, from the set of all genes, with the best predictive accuracy in a classification problem.

As Dougherty (2001) puts it: "The problem of error estimation impacts variable selection in a devilish way." As mentioned above, if our error estimate has a large variance, due to small-sample-size effects, there will be many overly optimistic small errors and consequently, many gene subsets that apparently yield excellent classification performance. It is important to be aware of this phenomenon and not be misled by an overabundance of "optimal" feature subsets.

Exhaustive search for the optimal feature subset is usually not feasible, owing to the large number of genes. There are various approaches that can be taken. For example, one could begin with no features and iteratively adjoin new promising ones (forward selection); or one could start with all features and iteratively remove those features that are believed to be uninformative or redundant (backward selection). There are also deterministic and stochastic feature subset search algorithms. The former type always produce the same feature subset on the same data set, whereas the latter type may produce different results on the same data but are more likely to find globally better

---

[10] Tenfold cross-validation is often employed and recommended.

solutions and not get stuck in local optima. Genetic algorithms and simulated annealing are two examples of stochastic search algorithms.

In order for us to determine the best feature subset, we must have some way of evaluating the performance and comparing candidate feature subsets. There are two main approaches, the filter and wrapper approaches (Kohavi and John, 1997). The *filter approach* uses some criterion (e.g., T-statistic) to evaluate the feature subset without regard for the actual classification rule. The *wrapper approach*, on the other hand, uses the classifier itself to evaluate a potential feature subset. The philosophy behind the latter approach is that the feature subset should also depend on the learning algorithm itself, as this algorithm will be the one that will ultimately be used for making predictions. Thus, for every candidate feature subset, the classifier is designed on it, its error is estimated, and the worth of the feature subset is evaluated on the basis of this error. It is commonly acknowledged that the wrapper approach attains predictive accuracy superior to that of the filter approach.

What, then, is the optimal size of the feature subset? If the feature subset is too small, our classifier will be overly biased toward the training set. Hence, its predictive accuracy will be sacrificed. If the feature subset is too large, it will probably include irrelevant (noisy) features that carry no predictive power. Consequently, the predictive accuracy will again be sacrificed. In the wrapper approach, since the size of the feature subset depends on the classifier error, sensible approaches to the design of classifiers and their error estimation, taking into account the sample size, as discussed in previous sections, will help ensure the correct choice of the feature subset size.

There is a substantially dangerous pitfall at the feature subset selection stage, known as the *selection bias* (Ambroise and McLachlan, 2002). The performance of the classification rule, for small sample sizes, is typically assessed using cross-validation methods, as discussed above. However, if the classifier is tested on samples that were already used for the feature selection step, or if the cross-validation-based error estimation of the classifier is 'within the loop' of feature selection, a selection bias occurs and the error estimate of the classifier is often overly optimistic. To overcome this problem, an external cross-validation loop should be undertaken, outside the feature subset selection step. As Ambroise and McLachlan (2002) point out, if the test set is used to estimate the classification error, then if it is also used for feature subset selection, a selection bias will be introduced. Consequently, the test set should never be used in the feature selection stage, although many studies unfortunately ignore this issue. However, as the sample sizes are typically quite small and we cannot afford to reserve a part of the data solely for testing, various selection bias corrections can be applied (Ambroise and McLachlan, 2002) in order to get a realistic estimate of the classification error.

REFERENCES

1. Ambroise C, McLachlan GJ. (2002) Selection bias in gene extraction on the basis of microarray gene-expression data. *Proc Natl Acad Sci USA* 99(10):6562–6.

2. Baggerly KA, Coombes KR, Hess KR, Stivers DN, Abruzzo LV, Zhang W. (2001) Identifying differentially expressed genes in cDNA microarray experiments. *J Comput Biol* 8(6):639–59.

3. Beissbarth T, Fellenberg K, Brors B, Arribas-Prat R, Boer J, Hauser NC, Scheideler M, Hoheisel JD, Schutz G, Poustka A, Vingron M. (2000) Processing and quality control of DNA array hybridization data. *Bioinformatics* 16(11):1014–22.

4. Black MA, Doerge RW. (2002) Calculation of the minimum number of replicate spots required for detection of significant gene expression fold change in microarray experiments. *Bioinformatics* 18(12):1609–16.

5. Bolstad BM, Irizarry RA, Astrand M, Speed TP. (2003) A comparison of normalization methods for high density oligonucleotide array data based on variance and bias. *Bioinformatics* 19(2):185–93.

6. Borg I, Groenen P. (1997) *Modern Multidimensional Scaling. Theory and Applications.* Springer Verlag, New York.

7. Brazma A, Vilo J. (2001) Gene expression data analysis. *Microbes Infect* 3(10):823–9.

8. Brazma A, Hingamp P, Quackenbush J, Sherlock G, Spellman P, Stoeckert C, Aach J, Ansorge W, Ball CA, Causton HC, Gaasterland T, Glenisson P, Holstege FC, Kim IF, Markowitz V, Matese JC, Parkinson H, Robinson A, Sarkans U, Schulze-Kremer S, Stewart J, Taylor R, Vilo J, Vingron M. (2001) Minimum information about a microarray experiment (MIAME)– toward standards for microarray data. *Nat Genet* 29(4):365–71.

9. Brun M, Sabbagh DL, Kim S, Dougherty ER. (2003) Corrected small-sample estimation of the Bayes error. *Bioinformatics* 19(8):944–51.

10. Chen Y, Dougherty ER, Bittner ML. (1997) Ratio-based decisions and the quantitative analysis of cDNA microarray images. *J Biomed Optics* 2(4):364–374.

11. Chen Y, Kamat V, Dougherty ER, Bittner ML, Meltzer PS, Trent JM. (2002) Ratio statistics of gene expression levels and applications to microarray data analysis. *Bioinformatics* 18(9):1207–1215.

12. Chu S, DeRisi J, Eisen M, Mulholland J, Botstein D, Brown PO, Herskowitz I. (1998) The transcriptional program of sporulation in budding yeast. *Science* 282(5389):699–705.

13. Dobbin K, Shih JH, Simon R. (2003) Statistical design of reverse dye microarrays. *Bioinformatics* 19(7):803–10.

14. Dougherty ER. (2001) Small sample issues for microarray-based classification. *Compar Funct Genomics* 2:28–34.

15. Dudley AM, Aach J, Steffen MA, Church GM. (2002) Measuring absolute expression with microarrays with a calibrated reference sample and an extended signal intensity range. *Proc Natl Acad Sci USA* 99(11):7554–9.

16. Dudoit S, Yang YH, Callow MJ, Speed TP. (2002) Statistical methods for identifying differentially expressed genes in replicated cDNA microarray experiments. *Statistica Sinica* 12:111–139.

17. Edwards D. (2003) Non-linear normalization and background correction in one-channel cDNA microarray studies. *Bioinformatics* 19(7):825–33.

18. Efron B, Tibshirani RJ. (1993) *An Introduction to the Bootstrap.* Chapman & Hall, New York.

19. Fang Y, Brass A, Hoyle DC, Hayes A, Bashein A, Oliver SG, Waddington D, Rattray M. (2003) *Nucleic Acids Res* 31(16):e96.

20. Finkelstein, D., Ewing, R., Gollub, J., Sterky, F., Cherry, J. M., Somerville, S. (2002) Microarray data quality analysis: lessons from the AFGC project. *Plant Mol Biol* 48(1–2):119–31.

21. Hoffmann R, Seidl T, Dugas M. (2002) Profound effect of normalization on detection of differentially expressed genes in oligonucleotide microarray data analysis. *Genome Biol* 3(7):research0033.

22. Hoyle DC, Rattray M, Jupp R, Brass A. (2002) Making sense of microarray data distributions. *Bioinformatics* 18(4):576–84.

23. Hsing T, Attoor S, Dougherty ER. (2003) Relation between permutation-test P values and classifier error estimates. *Machine Learning* 52(1–2):11–30.

24. Hu L, Cogdell D, Jia Y, Hamilton SR, Zhang W. (2002) Monitoring of microarray production with a common oligonucleotide and specificity with selected targets. *Biotechniques* 32:528–534.

25. Huber W, Von Heydebreck A, Sultmann H, Poustka A, Vingron M. (2002) Variance stabilization applied to microarray data calibration and to the quantification of differential expression. *Bioinformatics* 18 Suppl 1:S96–S104.

26. Ideker T, Thorsson V, Siegel AF, Hood LE. (2000) Testing for differentially-expressed genes by maximum-likelihood analysis of microarray data. *J Comput Biol* 7(6):805–17.

27. Jenssen TK, Langaas M, Kuo WP, Smith-Sørensen B, Myklebost O, Hovig E. (2002) Analysis of repeatability in spotted cDNA microarrays. *Nucleic Acids Res* 30(14):3235–44.

28. Kepler TB, Crosby L, Morgan KT. (2002) Normalization and analysis of DNA microarray data by self-consistency and local regression. *Genome Biol* 3(7):research0037.

29. Kerr MK, Martin M, Churchill GA. (2000) Analysis of variance for gene expression microarray data. *J Comput Biol* 7(6):819–37.

30. Kerr MK, Leiter EH, Picard L, Churchill GA. (2002a) Sources of variation in microarray experiments. In *Computational and Statistical Approaches to Genomics*, W. Zhang and I. Shmulevich, eds. Kluwer Academic Publishers, Norwell, MA.

31. Kerr MK, Afshari CA, Bennett L, Bushel P, Martinez J, Walker NJ, Churchill GA. (2002b) Statistical analysis of a gene expression microarray experiment with replication. *Statistica Sinica* 12:203–217.

32. Khan AH, Ossadtchi A, Leahy RM, Smith DJ. (2003) Error-correcting microarray design. *Genomics* 81(2):157–65.

33. Kohavi R, John GH. (1997) Wrappers for feature subset selection. *Artificial Intelligence* 97:273–324.

34. Kothapalli, T, Yoder, SJ, Mane, S, Loughran TP Jr. (2002) Microarray results: how accurate are they? *BMC Bioinformatics* 3(22).

35. Lee ML, Kuo FC, Whitmore GA, Sklar J. (2000) Importance of replication in microarray gene expression studies: statistical methods and evidence from repetitive cDNA hybridizations. *Proc Natl Acad Sci USA* 97(18):9834–9.

36. Lee PD, Sladek R, Greenwood CM, Hudson TJ. (2002) Control genes and variability: absence of ubiquitous reference transcripts in diverse mammalian expression studies. *Genome Res* 12(2):292–7.

37. Liang M, Briggs AG, Rute E, Greene AS, Cowley AW Jr. (2003) Quantitative assessment of the importance of dye switching and biological replication in cDNA microarray studies. *Physiol Genomics* 14(3):199–207.

38. Lönnstedt I, Speed T. (2002) Replicated microarray data. *Statistica Sinica* 12:31–46.

39. Love B, Rank DR, Penn SG, Jenkins DA, Thomas RS. (2002) A conditional density error model for the statistical analysis of microarray data. *Bioinformatics* 18(8):1064–72.

40. Novak JP, Sladek R, Hudson TJ. (2002) Characterization of variability in large-scale gene expression data: implications for study design. *Genomics* 79(1):104–13.

41. Pan W. (2002) A comparative review of statistical methods for discovering differentially expressed genes in replicated microarray experiments. *Bioinformatics* 18(4):546–54.

42. Pan W, Lin J, Le CT. (2002) How many replicates of arrays are required to detect gene expression changes in microarray experiments? A mixture model approach. *Genome Biol* 3(5):research0022.

43. Pan W. (2003) On the use of permutation in and the performance of a class of nonparametric methods to detect differential gene expression. *Bioinformatics* 19(11):1333–40.

44. Quackenbush J. (2002) Microarray data normalization and transformation. *Nat Genet* 32 Suppl:496–501.

45. Richmond CS, Glasner JD, Mau R, Jin H, Blattner FR. (1999) Genome-wide expression profiling in Escherichia coli K–12. *Nucleic Acids Res* 27(19):3821–35.

46. Rocke DM, Durbin B. (2001) A model for measurement error for gene expression arrays. *J Comput Biol* 8(6):557–69.

47. Schuchhardt J, Beule D, Malik A, Wolski E, Eickhoff H, Lehrach H, Herzel H. (2000) Normalization strategies for cDNA microarrays. *Nucleic Acids Res* 28(10):e47.

48. Shmulevich, I. (2003) Model Selection in Genomics. Guest Editorial in Toxicogenomics Section of *Environmental Health Perspectives*, 111(6):A328–A329.

49. Shmulevich I, Zhang W. (2002a) Binary Analysis and Optimization-Based Normalization of Gene Expression Data. *Bioinformatics* 18(4):555–65.

50. Shmulevich I, Hunt K, El-Naggar A, Taylor E, Ramdas L, Laborde P, Hess KR, Pollock R, Zhang W. (2002b) Tumor specific gene expression profiles in human leiomyosarcoma: an evaluation of intratumor heterogeneity. *Cancer* 94(7):2069–75.

51. Shmulevich I, Astola J, Cogdell D, Hamilton SR, Zhang W. (2003) Data extraction from composite oligonucleotide microarrays. *Nucleic Acids Res* 31(7):e36.

52. Somorjai RL, Dolenko B, Baumgartner R. (2003) Class prediction and discovery using gene microarray and proteomics mass spectroscopy data: curses, caveats, cautions. *Bioinformatics* 19(12):1484–91.

53. Spellman PT, Miller M, Stewart J, Troup C, Sarkans U, Chervitz S, Bernhart D, Sherlock G, Ball C, Lepage M, Swiatek M, Marks WL, Goncalves J, Markel S, Iordan D, Shojatalab M, Pizarro A, White J, Hubley R, Deutsch E, Senger M, Aronow BJ, Robinson A, Bassett D, Stoeckert CJ Jr, Brazma A. (2002) Design and implementation of microarray gene expression markup language (MAGE-ML). *Genome Biol* 3(9):research0046.

54. Tabus I, Rissanen J, Astola J. (2002) Normalized maximum likelihood models for Boolean regression with application to prediction and classification in genomics. In *Computational and Statistical Approaches to Genomics*, W. Zhang and I. Shmulevich, eds. Kluwer Academic Publishers, Norwell, MA.

55. Tsai CA, Chen YJ, Chen JJ. (2003) Testing for differentially expressed genes with microarray data. *Nucleic Acids Res* 31(9):e52.

56. Tseng GC, Oh MK, Rohlin L, Liao JC, Wong WH. (2001) Issues in cDNA microarray analysis: quality filtering, channel normalization, models of variations and assessment of gene effects. *Nucleic Acids Res* 29(12):2549–57.

57. Tsodikov A, Szabo A, Jones D. (2002) Adjustments and measures of differential expression for microarray data. *Bioinformatics* 18(2):251–60.

58. Troyanskaya O, Cantor M, Sherlock G, Brown P, Hastie T, Tibshirani R, Botstein D, Altman RB. (2001) Missing value estimation methods for DNA microarrays. *Bioinformatics* 17(6):520–5.

59. Troyanskaya OG, Garber ME, Brown PO, Botstein D, Altman RB. (2002) Nonparametric methods for identifying differentially expressed genes in microarray data. *Bioinformatics* 18(11):1454–61.

60. Tu Y, Stolovitzky G, Klein U. (2002) Quantitative noise analysis for gene expression microarray experiments. *Proc Natl Acad Sci USA* 99(22):14031–6.

61. Tusher VG, Tibshirani R, Chu G. (2001) Significance analysis of microarrays applied to the ionizing radiation response. *Proc Natl Acad Sci USA* 98(9):5116–21.

62. Wang X, Hessner MJ, Wu Y, Pati N, Ghosh S. (2003) Quantitative quality control in microarray experiments and the application in data filtering, normalization and false positive rate prediction. *Bioinformatics* 19(11):1341–7.

63. Wang Y, Lu J, Lee R, Gu Z, Clarke R. (2002) Iterative normalization of cDNA microarray data. *IEEE Trans Inform Technol Biomed* 6(1):29–37.

64. Wolkenhauer O, Möller-Levet C., Sanchez-Cabo F. (2002) The curse of normalization. *Comp Funct Genomics* 3:375–379.

65. Workman C, Jensen LJ, Jarmer H, Berka R, Gautier L, Nielser HB, Saxild HH, Nielsen C, Brunak S, Knudsen S. (2002) A new non-linear normalization method for reducing variability in DNA microarray experiments. *Genome Biol* 3(9):research0048.

66. Yang YH, Dudoit S, Luu P, Speed TP. (2001) Normalization for cDNA microarray data. *SPIE BiOS 2001*, San Jose, California, January 2001.

67. Yang YH, Dudoit S, Luu P, Lin DM, Peng V, Ngai J, Speed TP. (2002) Normalization for cDNA microarray data: a robust composite method addressing single and multiple slide systematic variation. *Nucleic Acids Res* 30(4):e15.

68. Yang IV, Chen E, Hasseman JP, Liang W, Frank BC, Wang S, Sharov V, Saeed AI, White J, Li J, Lee NH, Yeatman TJ, Quackenbush J. (2002) Within the fold: assessing differential expression measures and reproducibility in microarray assays. *Genome Biol* 3(11):research0062.

69. Zien A, Aigner T, Zimmer R, Lengauer T. (2001) Centralization: a new method for the normalization of gene expression data. *Bioinformatics* 17 Suppl 1:S323–31.

Index

DATE DUE

Demco, Inc. 38-293